Studies in Autonomic, Data-driven and Industrial Computing

The book series Studies in Autonomic, Data-driven and Industrial Computing (SADIC) aims at bringing together valuable and novel scientific contributions that address new theories and their real world applications related to autonomic, data-driven, and industrial computing. The area of research covered in the series includes theory and applications of parallel computing, cyber trust and security, grid computing, optical computing, distributed sensor networks, bioinformatics, fuzzy computing and uncertainty quantification, neurocomputing and deep learning, smart grids, data-driven power engineering, smart home informatics, machine learning, mobile computing, internet of things, privacy preserving computation, big data analytics, cloud computing, blockchain and edge computing, data-driven green computing, symbolic computing, swarm intelligence and evolutionary computing, intelligent systems for industry 4.0, as well as other pertinent methods for autonomic, data-driven, and industrial computing.

The series will publish monographs, edited volumes, textbooks and proceedings of important conferences, symposia and meetings in the field of autonomic, data-driven and industrial computing.

Bablu Kumar • Anshul Verma • Pradeepika Verma

Modern Kubernetes: From Core Concepts to Intelligent Autoscaling for Cloud Applications

Springer

Bablu Kumar
Department of Computer Science
Banaras Hindu University
Varanasi, Uttar Pradesh, India

Anshul Verma
Department of Computer Science
Banaras Hindu University
Varanasi, Uttar Pradesh, India

Pradeepika Verma
School of Computer Science Engineering & Technology
Bennett University
Greater Noida, Uttar Pradesh, India

ISSN 2730-6437 ISSN 2730-6445 (electronic)
Studies in Autonomic, Data-driven and Industrial Computing
ISBN 978-3-032-12971-0 ISBN 978-3-032-12972-7 (eBook)
https://doi.org/10.1007/978-3-032-12972-7

This Springer imprint is published by the registered company Springer Nature Switzerland AG
The registered company address is: Gewerbestrasse 11, 6330 Cham, Switzerland

Preface

Kubernetes has revolutionized the way modern applications are deployed, managed, and scaled by automating container orchestration, ensuring seamless performance, and optimizing resource utilization. With the rapid adoption of cloud-native technologies, Kubernetes has become a vital skill for IT professionals and a cornerstone of modern infrastructure. This book serves as a comprehensive guide to mastering Kubernetes—from beginner to advanced levels, providing both theoretical knowledge and practical insights to help readers navigate its complexities. This chapter introduces the fundamentals of Kubernetes, and its architecture, and offers a comparative overview of Docker Swarm and Kubernetes to establish a foundational understanding. It further explores the evolution of Kubernetes versions, highlighting key features and their roles in supporting stateful applications and ensuring data persistence. Core components such as Pods, Services, Deployments, ConfigMaps, and Networking are explained in detail. The chapter also delves into critical topics including reactive versus proactive scaling, intelligent autoscaling using the MAPE (Monitor-Analyze-Plan-Execute) framework, and resource management for cloud-native and edge applications. Additionally, it discusses approaches to evaluating system efficiency and performance metrics, offering valuable insights for both beginners and researchers.

- **Motivation Behind Writing This Book:** The growing demand for Kubernetes expertise inspired the creation of this book. Whether you are a beginner or an experienced professional, it simplifies complex concepts and offers a structured learning path. Through practical insights and real-world examples, it equips readers with the knowledge and skills needed to implement Kubernetes effectively. As Kubernetes has become essential for deploying and managing modern applications, this book covers both fundamental and advanced topics—including security, monitoring, and scaling. It also provides best practices and troubleshooting techniques to help readers efficiently manage Kubernetes clusters.

- **Book Structure and Learning Approach:** The book follows a structured approach, starting with fundamental concepts and progressively advancing to more complex topics. Each chapter includes practical examples, real-world scenarios, key takeaways, summaries, and review questions to reinforce learning and promote deeper exploration. This approach ensures that readers acquire both theoretical knowledge and the practical skills needed to apply Kubernetes concepts effectively in real-world environments.

Target Audience Designed for a diverse audience, this book is particularly well-suited for researchers focused on Docker and Kubernetes in cloud and edge computing, as well as for DevOps engineers, system administrators, software developers, and students eager to learn Kubernetes. Whether your goal is to automate deployments, manage pod scaling, implement reactive or proactive autoscaling, analyze performance metrics, optimize resource allocation, or enhance the efficiency of cloud-native applications, this book offers both foundational knowledge and advanced techniques to help you achieve those objectives. It caters to beginners seeking a structured introduction, as well as experienced professionals aiming to deepen their understanding of Kubernetes best practices.

The Journey of Writing This Book Writing this book has been a journey of continuous learning, research, and experimentation. Throughout the process, we have encountered challenges and breakthroughs that have deepened my understanding of Kubernetes. As an ever-evolving technology, staying updated with Kubernetes' latest trends and best practices has been both rewarding and challenging. By sharing these insights, we aim to help readers navigate Kubernetes adoption with confidence.

Final Thoughts As you embark on this journey through Kubernetes, we encourage you to explore its vast potential and apply the knowledge gained to real-world applications. Kubernetes is more than just a tool; it is a transformative technology shaping the future of cloud computing and modern software architecture. Mastering Kubernetes not only enhances your technical expertise but also opens doors to innovation, scalability, and efficiency in cloud-native environments. Let this book be your guide in navigating its complexities, gaining hands-on experience, and unlocking new possibilities in the ever-evolving cloud ecosystem. The future of cloud computing is being built today, so embrace the challenge and become a part of this revolution.

Acknowledgments We acknowledge the valuable online resources and tools that have supported the development of our content, many of which have been cited in this book. We also recognize the ongoing advancements in cloud-native technologies and express our sincere gratitude to the creators and platforms that provide open-access knowledge. The open-source community and Kubernetes practitioners continue to play a crucial role in the growth and evolution of the Kubernetes ecosystem. We gratefully acknowledge the support of the Institutions of Eminence (IoE) Scheme of Banaras Hindu University, Varanasi under Dev. Scheme No. 6031.

Varanasi, India — Bablu Kumar
Varanasi, India — Anshul Verma
Greater Noida, India — Pradeepika Verma

Competing Interests The authors have no competing interests to declare that are relevant to the content of this manuscript.

Contents

About the Authors

Bablu Kumar received an MCA degree in Computer Science and Engineering from Pondicherry University, Kalapet, Puducherry. He qualified for JRF in 2021 and is currently a research scholar under the supervision of Dr. Anshul Verma in the Department of Computer Science at the Institute of Science, Banaras Hindu University, Varanasi (India). His primary research interests include Cloud Computing, Edge Computing with Autoscaling and Resource Allocation, Docker, Kubernetes, Intelligent Systems, Distributed Systems, Deep Learning, and Federated Learning with Large Language Models (LLMs).

Anshul Verma received MTech and PhD degrees in Computer Science and Engineering from ABV-Indian Institute of Information Technology and Management Gwalior, India. He has done Post-doctorate from Indian Institute of Technology Kharagpur, India. Currently, he is serving as an Assistant Professor in the Department of Computer Science, Institute of Science, Banaras Hindu University, Varanasi, India. He has also served as a faculty member in the Computer Science and Engineering Department at Motilal Nehru National Institute of Technology (MNNIT) Allahabad and National Institute of Technology (NIT) Jamshedpur, India. His research interests include Cloud Computing, Distributed Systems, Mobile ad-hoc Networks, and Formal Verification. He is serving as an Associate Editor of the *Journal of Scientific Research* of the Banaras Hindu University and Editorial Board Member of Scientific Report journal of Springer.

Pradeepika Verma received her PhD degree in Computer Science and Engineering from the Indian Institute of Technology (ISM) Dhanbad, India. She has received MTech in Computer Science and Engineering from Banasthali University, Rajasthan, India. Currently, she is serving as an Assistant Professor at the School of Computer Science Engineering & Technology, Bennett University, Greater Noida, India. She has served as a Faculty Fellow in the Technical Innovation Hub at the Indian Institute of Technology, Patna, India and as a Post-Doctoral Fellow in the Department of Computer Science and Engineering at the Indian Institute of Technology (BHU), Varanasi, India. She has also worked as an Assistant Professor in the Department of Computer Science and Engineering at Pranveer Singh Institute of Technology, Kanpur, India, and as a Faculty Member in the Department of Computer Application at the Institute of Engineering and Technology, Lucknow, India. Her current research interests include Cloud Computing, Distributed Systems, Natural Language Processing, Optimization Approaches, and Mobile Ad-hoc Networks.

Chapter 1
Introduction of Kubernetes

Abstract This chapter introduces Kubernetes as a leading container orchestration platform that simplifies the deployment and management of microservices. It outlines the platform's key features and advantages over monolithic architectures, emphasizing scalability, resilience, and automation. The role of Kubernetes in enabling DevOps practices and supporting GitOps-driven deployment pipelines is explored, along with its challenges and benefits in real-world environments. Industry adoption and market impact are highlighted, showing how Kubernetes has become central to cloud-native development. The chapter also discusses career opportunities, certification paths, and salary trends, underlining the growing demand for Kubernetes expertise. Overall, it provides a concise foundation for understanding Kubernetes' technical, practical, and professional significance.

Keywords Kubernetes · Container orchestration · Microservices · DevOps · GitOps · Automation

Kubernetes originated as an evolution of Google's Borg system, which was designed to manage workloads at a massive scale within Google's internal infrastructure. Leveraging their experience with Borg, Google engineers developed Kubernetes as an open-source, community-driven project to provide enterprises with a robust container orchestration solution [1]. Recognizing its potential, Google donated Kubernetes to the Cloud Native Computing Foundation (CNCF) in 2014, ensuring its long-term development and widespread adoption [2]. Under CNCF's stewardship, Kubernetes has rapidly evolved, gaining strong industry support and continuous innovation from major technology companies, solidifying its role as the leading platform for container orchestration. Kubernetes is a powerful and flexible open-source container orchestration platform designed to automate containerised applications' deployment, scaling, and management [3]. By providing a consistent and unified platform, Kubernetes enables organizations to run applications seamlessly across various environments, including on-premises data centers, public clouds, and hybrid infrastructures. One of Kubernetes's core strengths lies in its

B. Kumar et al., *Modern Kubernetes: From Core Concepts to Intelligent Autoscaling for Cloud Applications*, Studies in Autonomic, Data-driven and Industrial Computing, https://doi.org/10.1007/978-3-032-12972-7_1

ability to allocate resources dynamically based on workload demands. It automates key operational tasks such as scaling applications, rolling out updates, and ensuring self-healing mechanisms, minimizing manual intervention and enhancing system reliability. These capabilities make Kubernetes an essential tool in modern DevOps practices, allowing teams to streamline application deployment, improve scalability, and optimize infrastructure utilization. As enterprises continue adopting cloud-native architectures, Kubernetes has emerged as the industry standard for managing containerized applications, offering automation, resilience, and efficiency in software deployment and operations. This chapter provides an overview of Kubernetes, highlighting its key characteristics, features, and role in orchestrating microservices. It discusses the challenges of monolithic architecture, industry adoption, and market impact. Additionally, it explores the integration of DevOps with Kubernetes, including benefits, challenges, and GitOps automation. Finally, it covers career opportunities, job trends, and certification programs for Kubernetes professionals.

1.1 Overview of Kubernetes

Kubernetes, often abbreviated as K8s, is an open-source platform designed to automate the deployment, scaling, and management of containerized applications [4]. This definition originates from Fig. 1.1 in the Kubernetes section. It was originally developed at Google based on their internal system called Borg and later donated to the Cloud Native Computing Foundation (CNCF) in 2014 [2]. Since then, Kubernetes has become the industry standard for container orchestration, widely adopted across enterprises, startups, and cloud providers. Kubernetes addresses many challenges of modern application deployment, particularly in cloud-native environments. It provides a unified platform that abstracts the complexities of infrastructure management, enabling developers to focus on building applications rather than handling resource allocation and system maintenance. To achieve this, Kubernetes is built on several key characteristics that define its overall capabilities. Figure 1.1 reflects the overview of Kubernetes, including its definition, control mechanisms, characteristics, and key features. It also highlights microservice orchestration, providing insights into Kubernetes' functionality.

1.1.1 Key Characteristics of Kubernetes

Kubernetes offers several essential features that make it a powerful platform for managing containerized applications efficiently.

- **Automation**: One of its most significant characteristics is automation. Kubernetes automates the deployment, scaling, and self-healing of applications, reducing the need for manual intervention. Through declarative configuration, users

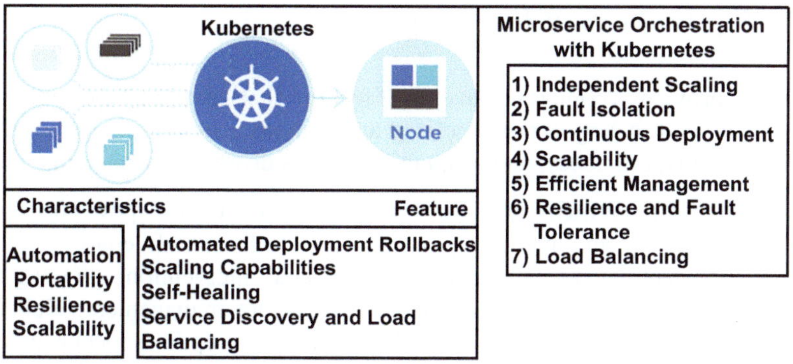

Fig. 1.1 Overview of Kubernetes

define the desired state of their applications, and Kubernetes ensures that the system continuously works to maintain that state. This automation extends to rolling updates, rollbacks, and service discovery, enhancing operational efficiency.

- **Portability**: Another key feature is portability. Kubernetes is designed to be platform-agnostic, meaning it can run in various environments, including on-premises data centers, public cloud platforms such as AWS, Google Cloud, and Azure, or even in hybrid cloud setups. This flexibility allows organizations to avoid vendor lock-in and deploy applications seamlessly across multiple infrastructures.
- **Resilience**: Resilience is another crucial advantage of Kubernetes. The system is built with fault tolerance in mind, ensuring that applications remain operational even in the face of failures. Kubernetes automatically detects and reschedules failed workloads to healthy nodes, maintaining application availability. Additionally, features like self-healing containers, automatic restarts, and replication help ensure minimal downtime.
- **Scalability**: Lastly, scalability is a fundamental strength of Kubernetes. It supports both horizontal and vertical scaling, allowing applications to adjust resources based on demand dynamically. Horizontal scaling involves adding or removing instances of a service, while vertical scaling adjusts the resources allocated to existing instances. Kubernetes' built-in scaling mechanisms, such as the Horizontal Pod Autoscaler (HPA) and Vertical Pod Autoscaler (VPA), help optimize resource utilization and improve performance.

By combining these characteristics, Kubernetes provides a highly automated, portable, resilient, and scalable environment for modern cloud-native applications.

1.1.2 Key Features of Kubernetes

Kubernetes provides a comprehensive set of features that streamline the deployment, management, and scaling of containerized applications. These capabilities enhance system reliability, efficiency, and adaptability across diverse environments

- **Automated Deployment and Rollbacks**: One of the most powerful features of Kubernetes is automated deployment and rollbacks. It enables seamless application deployment without requiring manual intervention, allowing teams to focus on development rather than infrastructure management. Kubernetes supports rolling updates, ensuring that application updates occur gradually without downtime. If an update introduces issues, its built-in rollback mechanism quickly restores the previous stable version, maintaining application reliability.
- **Scaling Capabilities**: Another essential feature is scaling capabilities. Kubernetes offers both horizontal and vertical scaling, making it adaptable to workload fluctuations. Horizontal scaling automatically adjusts the number of running instances (pods) based on CPU, memory, or custom-defined metrics, ensuring efficient resource utilization. Vertical scaling, on the other hand, optimizes resource allocation by dynamically adjusting the CPU and memory limits of individual containers.
- **Self-Healing**: Self-healing mechanisms further enhance Kubernetes' robustness. The platform continuously monitors container health, automatically restarting failed containers and rescheduling unresponsive pods to healthy nodes. If a node crashes or becomes unresponsive, Kubernetes redistributes workloads to maintain system stability.
- **Service Discovery and Load Balancing**: Kubernetes also simplifies service discovery and load balancing. It assigns DNS names to services, allowing applications to communicate seamlessly without requiring manual configuration of IP addresses. Additionally, its built-in load-balancing mechanisms efficiently distribute traffic across containers, preventing performance bottlenecks and ensuring high availability.

Finally, Kubernetes excels in multi-cloud and hybrid cloud support, making it a highly flexible solution for organizations deploying applications across diverse environments. It allows workloads to be distributed across on-premises data centers, public cloud platforms (such as AWS, Google Cloud, and Azure), and hybrid setups. This flexibility helps businesses avoid vendor lock-in and achieve greater deployment scalability. By integrating these characteristics and features, Kubernetes provides a comprehensive solution for orchestrating containerized applications, optimizing resource management, and ensuring high availability in modern cloud-native infrastructures. Modern application deployment has evolved from traditional monolithic architectures to more flexible microservices-based designs [3]. This shift enhances scalability, maintainability, and fault tolerance but also introduces operational complexities that necessitate robust container orchestration. Kubernetes plays a pivotal role in managing these complexities, providing automation, scalability, and resilience in cloud-native environments.

1.1.3 Challenges of Monolithic Architecture and the Need for Orchestration

In a monolithic architecture, all components of an application are tightly integrated and deployed as a single unit. While this approach simplifies initial development, it presents several challenges that hinder scalability, flexibility, and maintainability.

- Difficult to Scale: One major challenge is that monolithic applications are difficult to scale. Since all components are interconnected, scaling requires duplicating the entire system, even if only one component experiences high demand. This leads to inefficient resource utilization and increased operational costs.
- Single-Point Failures: Another significant issue is single-point failures. If any part of the application fails, the entire system can be impacted, reducing overall reliability and availability. This makes fault isolation difficult and increases downtime during failures.
- Complex Updates: Complex updates further complicate monolithic architectures. Any modification, whether small or large, requires redeploying the entire application. This not only increases deployment time but also raises the risk of introducing errors, making continuous integration and delivery (CI/CD) more challenging.
- Stack Rigidity: Additionally, technology stack rigidity limits flexibility. Since all components are built together, adopting new technologies, frameworks, or programming languages often requires a complete system overhaul, making innovation and modernization costly and time-consuming.
- Slow Development Cycles: Lastly, slow development cycles can hinder productivity. As the codebase grows, collaboration among development teams becomes challenging. Any change in one module may affect multiple parts of the application, requiring extensive testing and coordination before deployment.

To overcome these limitations, modern applications increasingly adopt microservices architectures, where an application is broken down into smaller, independent services that communicate via APIs. While this approach enhances flexibility and efficiency, managing and orchestrating multiple services across distributed environments becomes complex. This is where Kubernetes becomes essential, providing the automation, scalability, and orchestration needed to manage microservices efficiently.

1.1.4 The Role of Kubernetes in Microservices Orchestration

Microservices architecture enables modular application design by breaking down an application into independently running services [3]. Each microservice operates autonomously, bringing several advantages (Fig. 1.1):

- Independent Scaling: One of the key benefits is independent scaling. Unlike monolithic applications that require scaling the entire system, microservices allow individual services to scale based on demand. This optimizes resource utilization, reduces costs, and improves performance by allocating resources only where needed.
- Fault Isolation: Another advantage is fault isolation. Since each microservice runs separately, failures in one service do not necessarily impact the entire application. This enhances reliability and makes troubleshooting easier, as issues can be addressed at the service level without causing system-wide disruptions.
- Continuous Deployment: Additionally, continuous deployment is more efficient with microservices. Updates can be rolled out for specific services without affecting the rest of the system. This allows for faster development cycles, improved agility, and seamless integration of new features or bug fixes without downtime.

However, running multiple microservices in containers introduces new operational complexities, such as managing container lifecycles, handling failures, and ensuring efficient resource allocation. Kubernetes addresses these challenges through robust container orchestration, offering several key benefits:

- Scalability: Kubernetes enables auto-scaling by dynamically provisioning or terminating containers based on real-time demand. This ensures efficient resource usage while maintaining application performance.
- Efficient Management: Kubernetes provides a centralized control plane that automates deployment, monitoring, and lifecycle management of containers, reducing the need for manual intervention and operational overhead.
- Resilience and Fault Tolerance: Kubernetes enhances system reliability by detecting failures, restarting failed containers, rescheduling workloads on healthy nodes, and ensuring application uptime.
- Load Balancing: Kubernetes intelligently distributes network traffic across multiple pods to prevent bottlenecks and ensure even workload distribution, improving application responsiveness and availability.

By automating these critical aspects, Kubernetes empowers organizations to deploy, scale, and maintain their microservices efficiently in cloud-native environments. This makes Kubernetes an indispensable tool for modern software development, allowing businesses to focus on innovation rather than infrastructure management.

1.1.5 Industry Adoption with Real-World Use Cases

Kubernetes has been widely adopted across industries due to its ability to streamline application deployment, enhance scalability, and improve system resilience. Organizations across various domains leverage Kubernetes to manage large-scale, dynamic workloads efficiently. From e-commerce platforms to gaming applications, Kubernetes plays a crucial role in ensuring high availability, reliability, and automation in production environments. Several major companies have successfully integrated Kubernetes into their infrastructure to optimize operations and enhance performance:

- GitHub [5]: As a leading platform for software development and version control, GitHub relies on Kubernetes to automate its Continuous Integration/Continuous Deployment (CI/CD) pipelines. By leveraging Kubernetes, GitHub ensures seamless software updates, automated testing, and reliable deployment processes without manual intervention. This enables developers to push code changes frequently while maintaining stability and efficiency.
- Walmart [6]: In the retail sector, scalability is a critical requirement, particularly during seasonal events like Black Friday and Cyber Monday. Walmart deploys Kubernetes to manage its e-commerce infrastructure, allowing the platform to scale dynamically based on demand. This ensures a smooth shopping experience for millions of users, even during high-traffic periods. By automating resource allocation and load balancing, Kubernetes helps Walmart handle unpredictable surges while optimizing costs.
- Pokémon Go [7]: The global success of Pokémon Go required a highly scalable and resilient infrastructure capable of managing massive, unpredictable traffic spikes. Kubernetes played a pivotal role in enabling the game to scale dynamically, ensuring smooth gameplay for millions of concurrent players. By automatically distributing workloads and recovering from failures, Kubernetes helped maintain a seamless user experience despite fluctuating demands.

These real-world applications highlight Kubernetes's versatility in different industries, demonstrating its effectiveness in handling complex, large-scale workloads. As more businesses embrace cloud-native architectures, Kubernetes continues to be the go-to solution for container orchestration, driving innovation and operational efficiency across various sectors.

1.1.6 Market Impact

Kubernetes has significantly transformed the cloud computing landscape, becoming the de facto standard for container orchestration. According to surveys conducted by the Cloud Native Computing Foundation (CNCF), over 90% of enterprises have adopted Kubernetes, underscoring its widespread acceptance and dominance in modern IT infrastructure [2]. This high adoption rate reflects its critical role in enabling cloud-native computing and DevOps practices. As businesses increasingly shift towards microservices architectures and containerized applications, Kubernetes provides a unified platform for deploying, scaling, and managing workloads efficiently. Its automation capabilities reduce operational overhead, allowing organizations to streamline their software development lifecycle while ensuring high availability and performance.

Furthermore, Kubernetes has spurred innovation in multi-cloud and hybrid cloud strategies, enabling enterprises to deploy applications seamlessly across different cloud providers and on-premises environments. This flexibility helps organizations avoid vendor lock-in while optimizing costs and performance. The impact of Kubernetes extends beyond just IT operations; it has also influenced business agility and competitiveness. By facilitating rapid software deployment and continuous delivery, companies can accelerate product development cycles and respond swiftly to market demands. As a result, Kubernetes continues to be a driving force behind the modernization of enterprise infrastructure, reinforcing its position as a cornerstone of cloud-native computing.

1.2 DevOps and Its Role in Kubernetes

DevOps is a software development methodology that integrates development (Dev) and operations (Ops) teams to enhance collaboration, automation, and efficiency in software delivery [8]. It focuses on continuous integration and continuous deployment (CI/CD), infrastructure as code (IaC), monitoring, and automated testing. By breaking down silos between developers and IT operations, DevOps fosters a culture of agility, enabling faster and more reliable application releases. Kubernetes plays a critical role in modern DevOps practices by providing a scalable and automated platform for container orchestration. Since DevOps emphasizes automation and rapid deployment, Kubernetes aligns seamlessly with its principles by managing containerized applications efficiently. It automates tasks such as deployment, scaling, load balancing, and self-healing, reducing manual intervention and operational complexity. With Kubernetes, DevOps teams can implement rolling updates, canary deployments, and blue-green deployments, ensuring minimal downtime and smoother release cycles. In a typical DevOps workflow, Kubernetes is used alongside CI/CD pipelines to streamline application deployment. Developers push code changes to a version control system (e.g., Git), triggering automated testing and building processes. Once validated, Kubernetes deploys the new version of the application, ensuring high availability and rollback mechanisms in case of failures. Additionally, Kubernetes integrates with monitoring tools like Prometheus and Grafana, providing real-time insights into application performance and system health.

1.2.1 Benefits for DevOps

Kubernetes significantly enhances DevOps workflows by automating and optimizing application deployment, scaling, and management. It offers several key benefits that improve efficiency, flexibility, and cost-effectiveness in software development and operations.

- Speed: Kubernetes streamlines application deployment by automating the provisioning, scaling, and management of containers. This reduces manual effort, accelerates release cycles, and shortens the time-to-market for new features and updates. With Kubernetes, teams can deploy applications rapidly and roll out changes with minimal disruptions.
- Agility: Kubernetes is designed to support microservices architectures, allowing development teams to build and deploy independent services. Each microservice can be updated, tested, and scaled separately, enabling continuous integration and delivery (CI/CD). This modular approach enhances flexibility, making it easier to iterate, experiment, and refine applications without affecting the entire system.
- Cost Efficiency: Kubernetes optimizes resource usage by dynamically scaling workloads based on demand. It automatically allocates computing resources, ensuring that only necessary resources are used at any given time. This helps organizations reduce infrastructure costs by avoiding over-provisioning and making efficient use of available resources.
- Automation: One of Kubernetes' most powerful features is its ability to automate operational tasks, such as load balancing, self-healing, and rolling updates. This reduces the need for manual intervention, allowing DevOps teams to focus on innovation rather than infrastructure management. Kubernetes ensures high availability and reliability by automatically recovering from failures, further improving system resilience.

By combining DevOps principles with Kubernetes' orchestration capabilities, organizations can achieve a seamless, automated, and scalable application deployment process [8]. This synergy enables faster innovation, improved operational efficiency, and a more resilient infrastructure for modern cloud-native applications.

1.2.2 Challenges in Implementing DevOps with Kubernetes

While DevOps and Kubernetes together streamline application deployment, they also introduce certain challenges that organizations must address for successful implementation.

- Complexity in Setup and Management: Kubernetes has a steep learning curve, requiring teams to understand container orchestration concepts, networking, and resource management. Setting up and maintaining a Kubernetes cluster demands expertise, and misconfigurations can lead to security vulnerabilities or performance issues.

- Integration with Existing Workflows: Many organizations already have established CI/CD pipelines, monitoring tools, and security policies. Integrating Kubernetes into these existing DevOps workflows can be challenging, requiring modifications in infrastructure, tooling, and deployment strategies.
- Security and Compliance: Kubernetes environments introduce new security concerns, such as unauthorized access, misconfigured role-based access control (RBAC), and insecure container images. Ensuring compliance with industry regulations requires continuous monitoring, vulnerability scanning, and enforcing security best practices.
- Resource Optimization and Cost Management: Kubernetes allows dynamic scaling, but inefficient resource allocation can lead to excessive cloud costs. Organizations must implement proper monitoring and resource quotas to optimize workloads while balancing performance and budget constraints.
- Observability and Debugging: Debugging distributed applications running in Kubernetes can be more complex than traditional environments. Identifying performance bottlenecks and failures becomes difficult without proper observability [9] tools like Prometheus, Grafana, or Elasticsearch.

Despite these challenges, the combination of DevOps and Kubernetes offers significant advantages for modern cloud-native applications. By addressing these challenges through best practices, automation, and security policies, organizations can fully leverage Kubernetes in their DevOps workflows, achieving higher agility, scalability, and operational efficiency.

1.2.3 Kubernetes and GitOps: Enhancing Deployment Automation

GitOps is an emerging approach in DevOps that leverages Git as the single source of truth for managing Kubernetes infrastructure and application deployments [10]. By adopting GitOps, organizations can improve automation, traceability, and consistency in their deployment workflows.

- How GitOps Works with Kubernetes: In a GitOps workflow, all configuration files, including Kubernetes manifests, Helm charts, and deployment scripts, are stored in a Git repository. Instead of manually applying configurations, GitOps tools such as ArgoCD and Flux continuously monitor the repository for changes and automatically apply updates to the Kubernetes cluster. This ensures that the actual state of the system always matches the desired state defined in Git.
- Benefits of Using GitOps with Kubernetes

  - Declarative and Version-Controlled Infrastructure: Since all configurations are stored in Git, teams can easily track changes, roll back to previous versions, and maintain a clear audit history of deployments.
  - Automated and Consistent Deployments: GitOps eliminates manual intervention by automating deployments. When a developer pushes a change to Git, Kubernetes automatically applies the new configuration, ensuring a repeatable and consistent deployment process.
  - Improved Security and Compliance: By enforcing policy-driven deployments, GitOps ensures that only approved configurations are deployed. Role-based access control (RBAC) and code reviews further enhance security and compliance.
  - Enhanced Collaboration and Productivity: Developers can contribute to infrastructure and application configurations using Git's familiar workflow, making it easier to manage deployments across teams.

- Challenges of Implementing GitOps with Kubernetes

  - Initial Complexity: Setting up GitOps requires expertise in Git repositories, Kubernetes configurations, and continuous deployment (CD) tooling.
  - Managing Secrets: Sensitive information such as API keys and credentials must be securely handled using tools like Sealed Secrets or HashiCorp Vault.
  - Scaling and Performance: As deployments grow, managing multiple repositories and synchronizing changes across multiple clusters can become challenging.

GitOps enhances Kubernetes deployments by enabling automation, security, and version-controlled configurations. While it introduces some operational complexities, its benefits in maintaining consistency and reducing manual errors make it a powerful addition to DevOps workflows. Organizations looking to optimize their Kubernetes-based deployments can leverage GitOps to achieve better efficiency, reliability, and scalability.

1.2.4 Kubernetes Solution: Microservices Approach

Kubernetes plays a vital role in managing microservices-based architectures by breaking down applications into smaller, independent components [3]. This approach enhances scalability, resilience, and maintainability, making it ideal for modern cloud-native applications.

- Decoupling Application Components: Instead of deploying an application as a single monolithic unit, Kubernetes allows organizations to structure it into multiple microservices. Each microservice is responsible for a specific functionality, such as the frontend, backend, or database, and operates independently. This modular approach makes applications more flexible and easier to manage.

- Containerized Microservices: Each microservice runs in its own container, ensuring isolation and consistency across different environments. Kubernetes orchestrates these containers, handling their deployment, scaling, and communication. This improves portability and eliminates dependency conflicts between different services.
- Independent Scaling and Deployment: One of the biggest advantages of Kubernetes in a microservices architecture is its ability to scale and deploy services independently. If a specific microservice, such as the backend, experiences high demand, Kubernetes can scale that service without affecting other components. This optimizes resource utilization and ensures smooth performance.
- Fault Tolerance and Self-Healing: Kubernetes enhances application reliability by automatically detecting and recovering from failures. If a microservice crashes or becomes unresponsive, Kubernetes restarts it or reschedules it on a healthy node, minimizing downtime and maintaining application availability.

By leveraging Kubernetes for microservices, organizations can achieve greater agility, scalability, and fault tolerance while simplifying the management of complex distributed applications.

1.2.5 Getting Started with Kubernetes

For developers new to Kubernetes, there are essential tools that simplify cluster management and provide an efficient way to experiment with Kubernetes environments [4]. These tools help developers deploy, monitor, and scale containerized applications with ease.

- kubectl—the Kubernetes CLI Tool: kubectl is the primary command-line interface (CLI) tool used to interact with Kubernetes clusters [11]. It allows developers to deploy applications, inspect cluster resources, manage workloads, and troubleshoot issues. With kubectl, users can execute commands such as creating deployments, scaling services, and checking the health of running pods.
- Minikube—Local Kubernetes Testing: Minikube is a lightweight Kubernetes implementation designed for local testing and development [12]. It enables developers to create a single-node Kubernetes cluster on their local machine, allowing them to experiment with deployments without needing a full-scale cloud environment. Minikube provides an excellent starting point for learning Kubernetes concepts, running sample applications, and testing configurations before deploying them to a production environment.

By using these tools, developers can quickly gain hands-on experience with Kubernetes, understand how container orchestration works, and streamline application deployment processes.

Getting started with Kubernetes involves setting up a local development environment to explore its core functionalities. Developers can follow these initial steps to gain hands-on experience with container orchestration:

- Install Minikube on a Local Machine: Minikube provides a lightweight way to run a Kubernetes cluster locally. Installing it requires downloading the Minikube binary and ensuring dependencies like Docker, VirtualBox, or Hyper-V are set up. Minikube simplifies Kubernetes learning by creating a single-node cluster on a local machine.
- Start a Local Cluster with Minikube Start: Once Minikube is installed, developers can initiate a Kubernetes cluster using the command minikube start. This command sets up a fully functional Kubernetes environment, allowing users to experiment with deployments, scaling, and service management.
- Deploy Sample Applications: To understand Kubernetes operations, developers can deploy simple containerized applications [13]. By running kubectl apply -f <deployment.yaml>, they can launch and manage pods, services, and other Kubernetes resources. Testing with sample applications helps in grasping essential concepts such as load balancing, scaling, and rolling updates.
  Following these steps, developers can build a solid foundation in Kubernetes and gradually progress toward deploying and managing more complex microservices architectures.

1.3 Career Opportunities in Kubernetes

Kubernetes has become a critical skill in the modern IT landscape, significantly impacting career prospects in cloud computing, DevOps, and site reliability engineering (SRE). As organizations continue to adopt containerization and cloud-native technologies, professionals with Kubernetes expertise are in high demand.

1.3.1 Job Market and Salary Trends

The demand for Kubernetes professionals has surged, with companies actively seeking engineers skilled in container orchestration for roles such as DevOps engineers, site reliability engineers (SREs), and cloud infrastructure managers [14]. Kubernetes specialists command competitive salaries, with professionals in the United States earning an average of \$140,000 per year, while in India, salaries range between rupees 5–6 lakh per year. Given its central role in cloud-native development, Kubernetes expertise has become a valuable asset for IT professionals looking to advance their careers.

1.3.2 *Kubernetes Certification Programs*

To validate their expertise and stand out in the job market, professionals can pursue industry-recognized Kubernetes certifications:

- Certified Kubernetes Administrator (CKA): This certification demonstrates deep technical knowledge of Kubernetes operations, covering cluster installation, configuration, troubleshooting, and maintenance [15]. It is ideal for professionals managing Kubernetes infrastructure.
- Certified Kubernetes Application Developer (CKAD): Geared toward developers, this certification focuses on deploying, managing, and troubleshooting applications within Kubernetes environments. It emphasizes practical skills in containerized application development.
- Certified Kubernetes Security Specialist (CKS): Designed for security-focused professionals, this certification highlights best practices for securing Kubernetes workloads, addressing threats, and implementing compliance measures.

Earning these certifications not only enhances technical skills but also increases job prospects, making Kubernetes professionals highly sought after in the evolving cloud ecosystem.

1.4 Key Takeaways

- The chapter provides a comprehensive overview of Kubernetes, its core concepts, and its growing importance in modern cloud-native computing.
- It details Kubernetes' essential characteristics, including self-healing, automated rollouts and rollbacks, service discovery, and load balancing, which enable efficient container orchestration.
- The discussion highlights the challenges of monolithic architectures and explains how Kubernetes facilitates the shift to microservices through automated workload management, seamless scaling, and fault tolerance.
- Real-world examples illustrate Kubernetes' role in enhancing agility, optimizing resource utilization, and streamlining deployment processes, alongside its increasing adoption across various industries.
- Kubernetes strengthens continuous integration and continuous deployment (CI/CD) pipelines, supports Infrastructure-as-Code (IaC), and fosters better collaboration between development and operations teams.
- The chapter introduces GitOps as an extension of DevOps, showcasing how it enables declarative, version-controlled, and automated Kubernetes deployments while addressing challenges like secret management and scaling complexity.
- It outlines the installation steps and initial setup procedures for Kubernetes, providing a hands-on introduction for new users.

- Career Opportunities in Kubernetes: The chapter discusses the growing demand for Kubernetes expertise, salary trends, and the value of Kubernetes certification programs for professionals pursuing cloud-native roles [14].
- By covering Kubernetes' architecture, industry relevance, DevOps integration, and GitOps automation, the chapter equips readers with a solid foundation to understand Kubernetes' critical role in modern application deployment and management.

1.5 Summary

This chapter introduces Kubernetes as a critical platform for container orchestration and cloud-native application management, highlighting its role in automating deployment, scaling, and maintenance of containerized workloads. It begins by outlining Kubernetes' key characteristics, such as automated scaling, self-healing, and service discovery, which contribute to improved application performance, availability, and reliability. The discussion extends to monolithic vs. microservices architectures, emphasizing how Kubernetes simplifies microservices orchestration by ensuring efficient resource allocation, fault tolerance, and workload distribution. Real-world industry applications showcase how enterprises leverage Kubernetes to reduce infrastructure costs, streamline deployment pipelines, and enhance operational efficiency. Additionally, the chapter explores DevOps and GitOps integration, detailing how Kubernetes facilitates CI/CD workflows, automates infrastructure management, and supports declarative configuration practices. GitOps is introduced as a modern approach that extends DevOps principles by ensuring version-controlled, automated Kubernetes deployments, improving consistency and security while addressing challenges like secret management and workflow integration. The chapter also highlights career opportunities in Kubernetes, detailing the rising demand for Kubernetes expertise, lucrative salary trends, and the value of certifications such as CKA (Certified Kubernetes Administrator) and CKAD (Certified Kubernetes Application Developer) in advancing careers in cloud-native development and operations. Overall, this chapter provides a strong foundation for understanding Kubernetes, its advantages, industry impact, and career prospects, equipping readers with the knowledge to delve deeper into its technical aspects and real-world applications.

1.6 Review Questions

- What is Kubernetes and what primary problem does it solve in cloud computing?
- How does Kubernetes differ from traditional infrastructure management tools?
- List three defining characteristics of Kubernetes that make it suitable for modern applications.

- Explain how Kubernetes achieves “self-healing” capabilities.
- Describe how Kubernetes handles automated rollouts and rollbacks.
- What storage orchestration features does Kubernetes provide?
- What limitations of monolithic architectures led to the need for container orchestration?
- How does Kubernetes address the scaling challenges faced by monolithic applications?
- Explain how Kubernetes simplifies the deployment of microservices compared to manual methods.
- What Kubernetes feature is most critical for managing microservices communication?
- Name two major companies using Kubernetes and describe their implementation benefits.
- What industry trends have contributed to Kubernetes’ widespread adoption?
- How has Kubernetes influenced the cloud services market and vendor offerings?
- What three DevOps practices does Kubernetes enhance most significantly?
- Describe one challenge organizations face when integrating Kubernetes into DevOps workflows.

Chapter 2
Kubernetes Fundamentals

Abstract This chapter introduces the fundamental building blocks of Kubernetes and explains how they enable scalable and reliable application deployment. It begins with pods, the smallest deployable unit, and demonstrates their structure through a simple example. The lifecycle management of pods is addressed through deployments, while services are presented as the means to ensure network communication between components. The chapter also covers ConfigMaps and secrets, which provide efficient ways to manage configuration and sensitive data securely. YAML's role in Kubernetes is explained in detail, highlighting its structure, syntax, and purpose in defining resources. Finally, the concept of operators is introduced, showing how they extend Kubernetes' capabilities by automating complex operational tasks. Together, these concepts establish a strong foundation for understanding how Kubernetes organizes and manages applications in a cloud-native environment.

Keywords Kubernetes fundamentals · Pods · Deployments · Services · ConfigMaps · Secrets · YAML · Operators · Configuration management

Kubernetes is a powerful container orchestration platform that automates the deployment, scaling, and management of containerized applications. To fully leverage its capabilities, understanding its core concepts and abstractions is essential, as they form the foundation for orchestrating workloads in a cloud-native environment [16]. At the heart of Kubernetes architecture are the Master Node and Worker Nodes, which work together to manage containerized applications efficiently. This coordination ensures scalability, fault tolerance, and optimal resource utilization across the cluster, ultimately providing high availability, resilience, and seamless workload management in modern cloud-native environments. This chapter covers the fundamentals of Kubernetes, starting with core concepts such as Pods, Deployments, Services, ConfigMaps, and Secrets. It explains YAML configuration, its structure, and its importance in Kubernetes. Additionally, it introduces Kubernetes Operators, their setup, and troubleshooting for practical implementation.

B. Kumar et al., *Modern Kubernetes: From Core Concepts to Intelligent Autoscaling for Cloud Applications*, Studies in Autonomic, Data-driven and Industrial Computing, https://doi.org/10.1007/978-3-032-12972-7_2

2.1 Core Concepts Overview

Kubernetes simplifies container management by introducing key abstractions that help structure and organize workloads efficiently. These components work together to ensure high availability, scalability, and resilience for applications running in a Kubernetes cluster.

2.1.1 Pods: The Smallest Deployable Unit

A Pod is the smallest and most fundamental unit in Kubernetes [17]. It represents a single instance of a running process within the cluster. Unlike traditional containers, Pods can contain multiple tightly coupled containers that share network namespaces, storage volumes, and runtime environments. There are different types of pods:

- Single-Container Pods: A Single-Container Pod is the most common type of pod in Kubernetes, where each pod contains only one application container. This setup is ideal when an application component runs independently and does not require auxiliary services within the same pod. Examples include a web server, database instance, or API service, where each pod is responsible for hosting a single service that communicates with other components over the network. Since Kubernetes manages each pod separately, it ensures proper resource allocation, scaling, and fault tolerance for single-container deployments. These pods are easy to monitor, debug, and manage, making them the preferred choice for most cloud-native applications.
- Multi-Container Pods: A Multi-Container Pod consists of two or more containers that work closely together within the same pod [18]. Unlike single-container pods, these containers share the same network namespace, storage volumes, and inter-process communication mechanisms, allowing them to function as a cohesive unit. One common design pattern for multi-container pods is the sidecar pattern, where an additional container provides logging, monitoring, security, caching, or proxying capabilities to support the main application container. For example, a primary container running a web application might be accompanied by a sidecar container that collects logs and sends them to an external monitoring system. Multi-container pods enhance modularity and improve data processing, performance optimization, and security enforcement within Kubernetes-based applications.

Pods are ephemeral by nature, meaning they can be terminated, rescheduled, or replaced on different nodes in response to failures, scaling demands, or rolling updates [17]. This ensures high availability and fault tolerance within the cluster. A Pod is a fundamental building block in Kubernetes that abstracts the container runtime and provides a unified interface for deploying and managing applications.

It encapsulates one or more containers that share storage, networking, and execution specifications. For example, a pod can run a Nginx web server alongside a sidecar container responsible for logging. The pod lifecycle consists of several stages: it is created, scheduled to a node, runs until its task completes or fails, and is eventually terminated or restarted as needed, ensuring resilience and availability within the cluster. Deployment Example: This defines an Nginx deployment with three replicas, ensuring high availability.

2.1.2 Understanding Kubernetes Pod Syntax Through an Example

Kubernetes manifests are typically written in YAML format to define resources such as Pods, Deployments, and Services. The following example describes a simple Pod configuration for running an Nginx web server. Let's break down the key elements of the code snippet to understand its syntax and structure.

- API Version Specification: Every Kubernetes manifest begins by specifying the API version. In this example, v1 indicates that this Pod definition follows the first stable version of the Kubernetes API [19]. Different Kubernetes resources may require different API versions based on their availability and capabilities.

```
apiVersion: v1
```

- Defining the Resource Kind: The kind field specifies the type of Kubernetes object being defined. Here, the manifest describes a Pod, which is the smallest deployable unit in Kubernetes. Other resource types include Deployments, Services, ConfigMaps, and PersistentVolumes.

```
kind: Pod
```

- Metadata Section: The metadata section provides essential information about the resource, such as its name, labels, and annotations. In this case, the Pod is named nginx-pod, which helps uniquely identify it within the Kubernetes cluster.

```
metadata:
  name: nginx-pod
```

- Specifying Pod Specifications: The spec section defines the desired state of the Pod, including its containers, networking, and storage settings. The containers field lists one or more containers that will run inside the Pod.

```
spec:
  containers:
```

- Defining a Container: Each container inside a Pod is defined with specific parameters. The name: nginx field assigns a logical name to the container, which helps in logging and debugging. The image: nginx:latest field specifies the Docker image to be used. Here, nginx:latest pulls the most recent version of the Nginx image from Docker Hub.

```
    - name: nginx
      image: nginx:latest
```

- Exposing Container Ports: The ports field defines which ports should be exposed within the container. The containerPort: 80 field ensures that the container listens on port 80, which is the default HTTP port for web servers like Nginx.

```
      ports:
      - containerPort: 80
```

This Kubernetes Pod configuration creates a single-container Pod running the Nginx web server. The YAML manifest: Specifies the API version (v1). Defines the resource type as a Pod and Assigns a unique name (nginx-pod). Specifies an Nginx container with the latest image and Exposes port 80 for web traffic.

- Kubernetes Pod Example

```
apiVersion: v1
kind: Pod
metadata:
  name: nginx-pod
spec:
  containers:
  - name: nginx
    image: nginx:latest
    ports:
    - containerPort: 80
```

This structure ensures that Kubernetes can deploy, manage, and monitor the Nginx server efficiently within the cluster.

2.1.3 Deployments: Managing Pod Lifecycle

A Deployment is a higher-level abstraction that automates the creation, scaling, and management of Pods. It allows users to define the desired state of an application and ensures that the correct number of replicas are always running. Key Features of Deployments:

- Replica Management: Deployments ensure that a specified number of pod replicas are maintained, automatically scaling them up or down based on demand.

  - When a Deployment is created, Kubernetes ensures that the specified number of Pod replicas are always running.
  - If demand increases, the Deployment can scale up by adding more replicas, and distributing the load efficiently.
  - Conversely, when demand decreases, the Deployment scales down to free up resources and reduce costs.
  - The Horizontal Pod Autoscaler (HPA) can be configured to dynamically adjust the number of replicas based on CPU or memory usage. Example YAML for setting replica count:

```
apiVersion: apps/v1
kind: Deployment
metadata:
  name: my-app
spec:
  replicas: 3  # Ensures three instances of the application run
  selector:
    matchLabels:
      app: my-app
  template:
    metadata:
      labels:
        app: my-app
    spec:
      containers:
      - name: my-container
        image: my-app:latest
```

  In this example, Kubernetes ensures that three replicas of the my-app container are always running.

- Rolling Updates: A rolling update is a deployment strategy that gradually replaces old Pods with new ones without causing downtime [20]. This ensures continuous availability of the application during updates.
  - When a new version of an application is deployed, Kubernetes gradually replaces the existing Pods with new Pods.
  - This prevents application downtime by ensuring that some Pods remain operational while others are being updated.
  - The maxUnavailable and maxSurge parameters can be used to control the update process by specifying how many Pods can be taken down or created simultaneously. Example YAML for rolling update strategy:

```
strategy:
  type: RollingUpdate
  rollingUpdate:
    maxUnavailable: 1
    maxSurge: 1
```

With this strategy, Kubernetes replaces Pods one at a time, ensuring zero downtime for users.

- Self-Healing: If a Pod crashes or becomes unresponsive, the Deployment automatically replaces it, maintaining application availability. Kubernetes Deployments provide self-healing capabilities to maintain application stability and availability.
  - If a Pod crashes or becomes unresponsive, the Deployment controller detects the failure and automatically replaces the failed Pod with a new one.
  - Kubernetes constantly monitors the health of Pods and ensures that the desired number of replicas is always running.
  - Liveness and Readiness probes can be configured to define health checks, allowing Kubernetes to restart unhealthy containers. Example YAML for self-healing with a liveness probe:

```
livenessProbe:
  httpGet:
    path: /health
    port: 8080
  initialDelaySeconds: 3
  periodSeconds: 5
```

  In this case, Kubernetes checks the /health endpoint every 5 seconds, and if the container is unresponsive, it is automatically restarted.

These features make Kubernetes an essential tool for containerized application deployment, scalability, and reliability in modern cloud-native environments.

Deployments are commonly used for stateless applications, such as web servers or APIs, where persistent data is stored externally in databases or storage solutions. A Deployment is a higher-level Kubernetes object that manages the lifecycle of pods, defining how many replicas should run at any given time while providing features such as rolling updates, rollbacks, and self-healing. Rolling updates allow applications to be upgraded without downtime, while rollbacks enable reverting to a previous version if issues arise. If a pod crashes or becomes unresponsive, the Deployment ensures high availability by automatically replacing it. At its core, a ReplicaSet is responsible for maintaining a specific number of pod replicas, ensuring that the desired number of instances are always running. If a pod fails or is deleted, the ReplicaSet creates a new one to restore the desired state. Deployments internally manage ReplicaSets to facilitate scaling and seamless updates. Together, Deployments and ReplicaSets ensure fault tolerance, scalability, and efficient container management in Kubernetes.

- Kubernetes Deployment Example

```yaml
apiVersion: apps/v1
kind: Deployment
metadata:
  name: nginx-deployment
spec:
  replicas: 3
  selector:
    matchLabels:
      app: nginx
  template:
    metadata:
      labels:
        app: nginx
    spec:
      containers:
      - name: nginx
        image: nginx:latest
        ports:
        - containerPort: 80
```

In this example, the Deployment creates and manages three replicas of an Nginx pod. If any pod fails, Kubernetes ensures another one is created automatically. The Deployment uses a ReplicaSet internally to maintain the correct number of replicas and handle rolling updates efficiently.

2.1.4 Services: Enabling Network Communication

A Service in Kubernetes provides a stable network endpoint to connect Pods and ensure seamless communication [53]. Since Pods are ephemeral and their IP addresses change dynamically, Services offer a consistent way to access applications within the cluster. There are different types of kubernetes service:

- ClusterIP (Default): Provides internal communication between Pods within the cluster.
- NodePort: Exposes the Service on a static port on each node, making it accessible externally.
- LoadBalancer: Works with cloud providers to distribute traffic efficiently across multiple Pods.
- Headless Service: Used for direct Pod-to-Pod communication, commonly applied in stateful applications like databases.

Example: Exposing an Nginx Deployment Using a NodePort Service

The following command exposes the `nginx-deployment` as a `NodePort` service, making it accessible on a fixed port from any node in the cluster. Kubernetes dynamically assigns a port in the range 30000-32767, allowing external users to reach the application.

```
kubectl expose deployment nginx-deployment --type=NodePort --
    port=80
```

By decoupling Pods from their networking, Services enable scalable and fault-tolerant application deployments.

2.1.5 *ConfigMaps and Secrets: Managing Configuration and Sensitive Data*

Kubernetes enables applications to externalize their configuration and sensitive data using ConfigMaps and Secrets, enhancing security, flexibility, and portability [45]. ConfigMaps store non-sensitive configuration data such as environment variables, file paths, and command-line arguments, allowing applications to be dynamically configured without modifying the container image. In contrast, Secrets securely store sensitive information like API keys, database credentials, and TLS certificates, ensuring encryption and strict access controls. Both ConfigMaps and Secrets can be mounted as volumes or injected as environment variables, enabling applications to adapt to configuration changes without requiring redeployment. ConfigMaps store non-sensitive configuration data, such as environment variables or application settings.

```
apiVersion: v1
kind: ConfigMap
metadata:
  name: app-config
data:
  database_url: mysql://localhost
```

Secrets store sensitive data, such as API keys and passwords, in an encrypted format.

```
apiVersion: v1
kind: Secret
metadata:
  name: my-secret
stringData:
  password: my-secure-password
```

The command kubectl expose deployment nginx-deployment –type=NodePort –port=80 is used in Kubernetes to create a Service that allows external access to an application running inside the cluster. Kubernetes pods are ephemeral, and their IP addresses change frequently, making direct access unreliable. By exposing a Deployment as a Service, this command ensures that the application remains accessible even if the underlying pods restart or move to different nodes. Breaking down the command, kubectl expose is the primary command used to create a Service for an existing resource, such as a Deployment, Pod, or ReplicaSet. The argument deployment nginx-deployment specifies that the Service should be created for the nginx-deployment, ensuring all pods associated with this deployment are accessible through a single, stable endpoint. The flag –type=NodePort designates the service as a NodePort, which means Kubernetes will assign a static port in the range of 30000-32767 on every node in the cluster. This allows external users to access the application by using the <NodeIP>:<NodePort> format. The –port=80 option specifies that the Service will forward traffic to port 80 inside the container, which is commonly used for web applications. Once this command is executed, Kubernetes automatically assigns a port within the NodePort range, which can be checked using kubectl get services. Users can then access the application externally by entering the NodeIP: AssignedPort in a browser or API client. While NodePort Services are useful for development and testing, they are not recommended for production environments due to security and scalability limitations. Instead, LoadBalancer or Ingress resources are preferred for handling external traffic in a more efficient and secure manner.

These core concepts—Pods, Deployments, Services, ConfigMaps, and Secrets—form the foundation of Kubernetes' orchestration capabilities. They work together to ensure scalability, fault tolerance, and efficient resource utilization, making Kubernetes an ideal platform for managing containerized applications in modern cloud-native environments.

2.2 YAML in Kubernetes

YAML (Yet Another Markup Language) is a widely used format for defining Kubernetes objects. It provides a human-readable structure that makes it easy to create, modify, and manage Kubernetes configurations. YAML follows a structured format using key-value pairs, lists, and maps, making it an efficient choice for configuration files in Kubernetes deployments.

2.2.1 Why Use YAML in Kubernetes?

YAML offers several advantages that make it a preferred choice for defining Kubernetes objects. One of the primary benefits is readability. YAML is designed to be human-friendly, using indentation to structure data rather than relying on complex syntax or brackets [23]. This makes it easier to read and write compared to alternatives like JSON or XML. The clear structure allows developers to quickly understand and modify configurations without extensive technical knowledge. Another significant advantage is portability. YAML is widely compatible with multiple programming languages, including Python and Java, making it a versatile option for various development and deployment environments. Since YAML files are plain text, they can be easily shared and used across different platforms without compatibility issues. This ensures that Kubernetes configurations remain flexible and adaptable to different workflows. Additionally, YAML enhances maintainability by allowing configuration files to be version-controlled using Git [24]. This makes it easy to track changes, revert to previous versions, and collaborate effectively within development teams. By storing YAML files in repositories, organizations can maintain consistency across deployments, reducing errors and ensuring smooth updates to Kubernetes resources.

2.2.2 Breakdown of the YAML Configuration

This YAML file defines a Kubernetes Pod, which is the smallest deployable unit in a Kubernetes cluster. The key components of the file are explained below:

- **apiVersion: v1**
  The `apiVersion` field specifies the Kubernetes API version being used. In this case, `v1` refers to the core API version supporting Pods.
- **kind: Pod**
  The `kind` field indicates the type of Kubernetes object being created. Here, it specifies that the object is a Pod.
- **metadata**
  The `metadata` section contains metadata information about the Pod, such as its name. The `name: my-pod` field assigns the Pod a unique identifier within the cluster.
- **spec**
  The `spec` section defines the Pod's specifications, including the list of containers it will run.
- **containers**
  The `containers` field specifies a list of container definitions. Each container is defined within a list item (preceded by a hyphen `-`).
- **name: nginx**
  This assigns a name to the container within the Pod, making it easier to reference within Kubernetes commands.
- **image: nginx**
  This field specifies the container image that should be used. The example pulls the official Nginx image from Docker Hub by default, allowing the Pod to run an Nginx web server.

1. Applying the YAML Configuration: To deploy the Nginx Pod, the YAML file must be applied using the `kubectl` command:

```
kubectl apply -f my-pod.yaml
```

   Once applied, Kubernetes schedules the Pod on an available node and ensures the container is running. Users can check the status of the Pod using:

```
kubectl get pods
```

2. Extending Kubernetes YAML Configurations: The basic Pod definition shown above serves as a foundation for more complex Kubernetes workloads. Additional configurations such as resource limits, environment variables, and volume mounts can be specified to optimize application deployment. For example, adding resource limits to control CPU and memory usage:

```
spec:
containers:

name: nginx
image: nginx
resources:
limits:
memory: "512Mi"
cpu: "500m"
```

   This configuration ensures that the Nginx container does not exceed 512Mi of memory and 500m (0.5 cores) of CPU resources.

2.2.3 *Structure of YAML in Kubernetes*

Kubernetes YAML manifests define objects using a specific structure. The essential components of a YAML file in Kubernetes include:

- **Key-Value Pairs**: Represented as `key: value`, these are used to define attributes such as the API version, object kind, and metadata.
- **Lists (Arrays)**: Defined using a hyphen (`-`) before each item. Lists are used when multiple values need to be specified, such as container definitions within a pod.
- **Maps (Dictionaries)**: Represent nested structures where a key contains another key-value pair inside `{}` or indented blocks. Maps are useful for defining more complex attributes like container specifications.

Example: YAML for a Kubernetes Pod Below is an example YAML file that defines a Kubernetes Pod running an Nginx container:

```yaml
apiVersion: v1
kind: Pod
metadata:
  name: my-pod
spec:
  containers:
  - name: nginx
    image: nginx
```

This makes it easier to read and distinguish different elements within the YAML configuration. The YAML content within this command defines a Kubernetes Pod, which is the smallest deployable unit in a Kubernetes cluster. The first line, apiVersion: v1, specifies the API version being used. In this case, v1 represents the core API version that supports Pods. The kind: Pod line indicates that the object being created is a Pod, which is essentially a wrapper around one or more containers running within a Kubernetes cluster. Next, the metadata section contains important details about the Pod, such as its name. In this example, name: my-pod assigns the name my-pod to the Pod, making it easier to reference within the cluster. The spec section follows, which defines the specifications of the Pod, including the containers that it should run. Within the containers section, a list of containers is specified. In this example, a single container is being defined with the name nginx using name: nginx. The image: nginx line specifies the container image that should be used. By default, this pulls the official Nginx image from Docker Hub, allowing the Pod to run an Nginx web server. This YAML configuration is used to deploy an Nginx container within a Kubernetes cluster. Once applied using the command kubectl apply -f <filename>.yaml, Kubernetes schedules the Pod on an available node and ensures the container is running. This basic example serves as a starting point for defining more complex Kubernetes workloads, where additional configurations such as resource limits, environment variables, and volume mounts can be added to enhance the deployment. Overall, YAML is an essential component of Kubernetes, providing a structured way to define and manage cluster resources. Understanding the structure and syntax of YAML manifests enables users to efficiently deploy and configure applications within Kubernetes environments. The ability to modify and extend these configurations allows for scalable, reliable, and well-managed cloud-native applications.

2.3 Operators

Kubernetes Operators enhance automation by extending the platform's capabilities, enabling intelligent management of complex applications. They encapsulate operational knowledge into software, allowing seamless deployment, scaling, and maintenance of stateful workloads within a Kubernetes cluster.

2.3.1 Operators Overview

Kubernetes Operators extend the platform's functionality by automating the management of complex applications [25]. They go beyond Kubernetes' built-in controllers by embedding human operational knowledge into software, enabling automated scaling, monitoring, and self-healing of applications. This makes Operators essential for running stateful applications efficiently within a Kubernetes cluster. A Kubernetes Operator is essentially a custom controller designed to manage applications that require ongoing maintenance [26]. Unlike default Kubernetes controllers that handle basic deployment and scaling, Operators introduce intelligence by automating application-specific tasks. They achieve this by using Custom Resource Definitions (CRDs) to extend the Kubernetes API, allowing users to define new resource types that cater to specific application needs. Operators function by continuously monitoring the application's state and executing necessary actions to maintain its desired configuration. For example, a PostgreSQL Operator can automate tasks such as database backups, resource scaling, failover handling, and version upgrades. Instead of relying on a database administrator for these tasks, an Operator ensures that they are performed automatically, improving efficiency and reducing the risk of errors. The key benefits of Kubernetes Operators include enhanced automation, as they eliminate the need for manual intervention in routine tasks. They also provide consistency by enforcing best practices, ensuring that applications are deployed and managed following predefined policies. Moreover, Operators contribute to scalability, dynamically adjusting resources based on demand, and resilience, improving fault tolerance by detecting and addressing failures proactively.

2.3.2 Operator Setup with Kubernetes

To establish a connection with Kubernetes using an Operator, follow these steps:

- Install the Operator SDK (if not already installed): Operators are typically built using the Operator SDK, which helps in developing and managing Operators [27]. Install it using the following command:

```
curl -LO https://github.com/operator-framework/operator-sdk/
    releases/latest/download/operator-sdk_linux_amd64
chmod +x operator-sdk_linux_amd64
sudo mv operator-sdk_linux_amd64 /usr/local/bin/operator-sdk
```

- Deploy the Operator in Kubernetes: Operators use Custom Resource Definitions (CRDs) to extend the Kubernetes API. To deploy an Operator, apply the necessary CRDs and Operator YAML files:

```
kubectl apply -f operator.yaml
kubectl apply -f custom-resource-definition.yaml
```

- Verify the Operator Installation: Check if the Operator is running by listing the pods in the namespace where it is deployed:

```
kubectl get pods -n operator-namespace
```

- Create a Custom Resource (CR) to Manage an Application: Once the Operator is deployed, create a CR that defines the desired state of the application it manages. Example (for a PostgreSQL Operator):

```
kubectl apply -f postgresql-cr.yaml
```

- Verify that the Operator is Managing the Resource: Check if the Operator has successfully created and is managing the resource:

```
kubectl get customresourcedefinitions
kubectl get postgresql -n default
```

  This process ensures that the Kubernetes Operator is connected and functioning correctly within the cluster.

In contrast, Kubernetes Operators play a crucial role in simplifying the management of complex applications. By integrating automation and intelligence into Kubernetes, they enhance reliability, scalability, and maintainability. Operators are particularly beneficial for managing stateful workloads such as databases, messaging systems, and distributed applications, making Kubernetes more adaptable for cloud-native environments.

2.3.3 *Practical and Troubleshooting*

To reinforce the concept of Kubernetes Operators, let's perform a hands-on exercise by deploying an Operator and managing an application with it [28].

- Deploying an Operator in Kubernetes: Operators rely on Custom Resource Definitions (CRDs) to extend Kubernetes capabilities [11]. First, we apply the CRD and Operator manifest using:

```
kubectl apply -f pod.yaml
```

  This command registers the Operator in the cluster and enables it to manage its defined custom resources.

- Verifying Operator Deployment: Once deployed, we can check the Operator's status using:

```
kubectl get pods -n <operator-namespace>
```

If the Operator is running correctly, it should appear as a running pod in the specified namespace.

- Creating a Custom Resource (CR) Managed by the Operator: Operators use Custom Resources (CRs) to define the applications they manage. We create a resource instance using:

```
kubectl apply -f custom-resource.yaml
```

This instructs the Operator to manage and configure the specified application automatically.

- Troubleshooting Operator Issues: If the Operator fails to work as expected, we can inspect its logs with:

```
kubectl logs <operator-pod-name> -n <operator-
    namespace>
```

For further debugging, we describe the Operator pod to check events, errors, or configuration details:

```
kubectl describe pod <operator-pod-name> -n <
    operator-namespace>
```

Additionally, To list all related resources in the operator's namespace:

```
kubectl get all -n <operator-namespace>
```

When working with Kubernetes, debugging issues is a crucial skill to diagnose and resolve problems efficiently [9]. To check the logs of a specific pod and analyze its output, we use:

```
kubectl logs <pod-name>
```

If further inspection is needed, we can get detailed information about a pod, including events, resource usage, and potential failures, by running:

```
kubectl describe pod <pod-name>
```

For a broader overview of all resources within a particular namespace, the following command provides a complete list of pods, services, deployments, and other components:

```
kubectl get all -n <namespace>
```

By mastering these essential Kubernetes commands, developers and system administrators can effectively deploy, monitor, and troubleshoot cloud-native applications, ensuring optimal performance and reliability.

2.4 Key Takeaways

- Introduction to Kubernetes Fundamentals: The chapter provides a comprehensive overview of the essential components that form the backbone of a Kubernetes cluster.
- Pods: Describes Pods as the smallest deployable unit that encapsulates applications and their dependencies, ensuring efficient container management.
- Deployments: Introduces Deployments as a mechanism for managing pod lifecycles, scaling, and automating updates while maintaining application stability.
- Services: Explains how Services facilitate communication between pods and enable seamless networking within a Kubernetes cluster.
- ConfigMaps and Secrets: Highlights their role in managing configurations and sensitive data, such as API keys and passwords, without embedding them directly into application code.
- YAML for Kubernetes Automation: Covers YAML as the declarative language for defining Kubernetes resources, explaining its structure, usage in configuring deployments, and importance in automation.
- Operators in Kubernetes: Discusses Operators as an advanced feature that automates complex application lifecycle management, particularly for stateful applications.
- Troubleshooting Techniques: Provides practical methods to diagnose and resolve common Kubernetes issues, ensuring smooth cluster operations.

2.5 Summary

This chapter lays the foundation for understanding Kubernetes by introducing its core concepts and configuration mechanisms. It begins with an overview of Pods, which act as the primary unit of deployment, encapsulating one or more containers that share storage and networking. It then discusses Deployments, which provide a way to define and automate the scaling, rolling updates, and rollback of applications. Services play a crucial role in enabling communication between different components of the cluster, ensuring that workloads remain accessible despite pod restarts or rescheduling. ConfigMaps and Secrets offer a flexible and secure method for managing configuration data and sensitive information separately from container images, enhancing security and maintainability. The chapter then transitions to YAML, explaining its role in defining Kubernetes objects declaratively. It covers the structure of YAML files, including key-value pairs, indentation rules, and multi-document support, demonstrating how YAML simplifies Kubernetes resource management. The discussion further explores Operators, which extend Kubernetes' automation capabilities beyond basic deployment management, allowing users to define custom controllers for handling application-specific tasks. To ensure practical knowledge, the chapter concludes with troubleshooting techniques, providing insights into diagnosing and resolving Kubernetes issues, such as pod failures, misconfigurations, and networking problems. This comprehensive coverage equips readers with the foundational knowledge needed to deploy, configure, and manage Kubernetes workloads efficiently.

2.6 Review Questions

- What are the four core components of Kubernetes architecture?
- Explain the relationship between a Pod and a Container in Kubernetes.
- Why are Pods considered the smallest deployable unit in Kubernetes?
- Describe a real-world scenario where multi-container Pods would be beneficial.
- How do Deployments improve Pod management compared to manual creation?
- What problem do Kubernetes Services solve in network communication?
- Compare ClusterIP, NodePort, and LoadBalancer Service types with use cases.
- When would you choose ConfigMaps over environment variables for configuration?
- What security best practices should be followed when using Kubernetes Secrets?
- Why is YAML the preferred format for Kubernetes manifests over JSON?
- Identify three mandatory fields in every Kubernetes YAML manifest.
- How does Kubernetes handle YAML validation errors during deployment?
- What are Operators in Kubernetes and how do they extend functionality?
- Outline the step-by-step process to install an Operator using Helm.
- What troubleshooting command would you use if an Operator-managed Pod fails to start?

Chapter 3
Kubernetes Architecture

Abstract This chapter explores the architecture of Kubernetes, focusing on how its core components work together to manage containerized applications. It begins with an overview of the master node and worker node components, explaining their distinct responsibilities in scheduling, execution, and cluster management. The discussion extends to scalability, including strategies for handling large-scale clusters and example commands for cluster configuration. Special attention is given to etcd, the distributed key-value store, where the Raft consensus algorithm ensures consistency; practical examples show how pod metadata and other data are stored and retrieved. Cluster communication and high availability are also addressed, with sample commands illustrating how nodes maintain reliable interactions. The chapter concludes with a hands-on guide to setting up a Kubernetes cluster, integrating Docker, and troubleshooting common node issues. Together, these topics provide a clear understanding of the internal design, scalability, and operational resilience of Kubernetes clusters.

Keywords Master node · Worker node · etcd · Cluster communication · Docker integration · Troubleshooting

Kubernetes is a powerful container orchestration system designed for managing large-scale applications. This chapter provides an overview of Kubernetes architecture, detailing the roles of master and worker nodes, scalability considerations, and key components like etcd [29]. It also covers cluster communication, high availability, and practical examples for setting up and troubleshooting a Kubernetes cluster (See Fig. 3.1). This chapter explores the architecture of Kubernetes, detailing the roles of master and worker node components. It discusses scalability limits, strategies for managing large-scale clusters, and commands for scaling. Additionally, it provides an in-depth look at etcd, covering the Raft consensus algorithm, metadata storage, and data retrieval. The chapter also explains cluster communication, ensuring high availability, and includes example commands. Finally, it presents a practical guide on setting up a Kubernetes cluster, integrating Docker, and troubleshooting node issues.

B. Kumar et al., *Modern Kubernetes: From Core Concepts to Intelligent Autoscaling for Cloud Applications*, Studies in Autonomic, Data-driven and Industrial Computing, https://doi.org/10.1007/978-3-032-12972-7_3

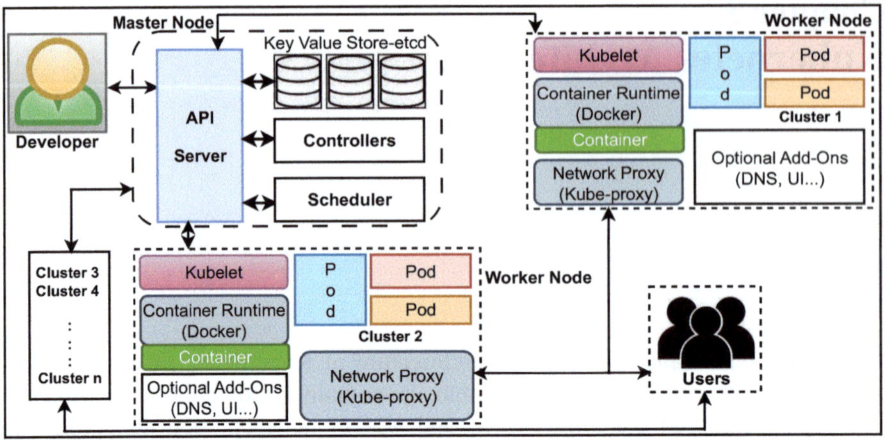

Fig. 3.1 Architecture of Kubernetes

3.1 Overview of Kubernetes Architecture

Kubernetes is a distributed container orchestration system designed to automate the deployment, scaling, and management of application containers across a cluster of machines [29]. It follows a master-worker architecture, where the master node oversees cluster operations while worker nodes execute containerized applications (Fig. 3.1). This separation of roles ensures efficient resource management, high availability, and fault tolerance in cloud-native environments. One of the key strengths of Kubernetes is its declarative configuration model. Users define the desired state of their applications, and Kubernetes automatically ensures that the system matches this state. This approach simplifies management and enhances reliability by reducing manual intervention. Kubernetes also provides automated management features, such as self-healing, auto-scaling, and rolling updates. Self-healing ensures that failed containers are restarted automatically, while auto-scaling adjusts resources dynamically based on demand. Rolling updates enable seamless application upgrades without downtime, ensuring continuous service availability. Another essential aspect of Kubernetes is service discovery and load balancing, which optimize traffic routing between applications. Kubernetes dynamically assigns network addresses and distributes traffic efficiently, ensuring seamless communication between microservices and improving overall system performance. Understanding its architecture is essential to effectively deploying and managing applications at scale.

3.2 Master Node Components in Kubernetes

The master node is the control center of a Kubernetes cluster, responsible for managing the cluster's state and ensuring that the desired application state matches the actual system state. It consists of several key components that work together to maintain orchestration, scheduling, and overall system reliability [30].

1. **API Server (kube-apiserver):** The API Server serves as the front-end for the Kubernetes cluster, acting as an interface between users, command-line tools (e.g., kubectl), and internal components. It processes REST API requests, validates them, and updates the cluster state stored in etcd. The API Server is the only component that directly interacts with etcd, ensuring secure and controlled access to cluster data. The Node Controller is responsible for monitoring the health of worker nodes in the Kubernetes cluster. If a node fails or becomes unresponsive, the Node Controller detects the issue and takes corrective actions, such as rescheduling pods to healthy nodes. This helps maintain the stability and availability of applications running in the cluster. The Replication Controller ensures that the correct number of pod replicas are always running based on the desired configuration. If a pod fails, the controller automatically creates a new one to replace it. Similarly, if more replicas are needed due to increased demand, the number of pods is scaled up to meet the workload requirements. The Endpoints Controller plays a crucial role in service discovery by keeping track of the relationship between services and their corresponding pods. When pods start, stop, or move to different nodes, the Endpoints Controller updates endpoint records to ensure that network requests are correctly routed to the active pods. This ensures smooth communication between services within the cluster.

```
kubectl get pods --all-namespaces
```

2. **Scheduler (kube-scheduler):** The Scheduler is responsible for assigning pods to worker nodes based on resource availability, constraints, and policies [31]. It ensures the optimal distribution of workloads while considering factors such as CPU, memory, and storage requirements. The Kubernetes Scheduler is responsible for deciding which worker node a pod should run on based on available resources. When a new pod is created, the scheduler evaluates factors such as CPU, memory, and storage availability across nodes to ensure efficient workload distribution. This helps optimize resource utilization and maintain cluster performance. To improve pod placement, the scheduler applies affinity and anti-affinity rules. Affinity rules allow related pods to be placed together for better communication, while anti-affinity rules ensure that certain pods are scheduled on separate nodes to enhance fault tolerance and prevent failures from affecting multiple instances of an application. Another key function of the scheduler is preventing resource contention by avoiding the placement of pods on overutilized nodes. If a node is already running at high capacity, the scheduler directs new workloads to less crowded nodes, ensuring a balanced distribution of

resources and preventing performance degradation. Example Scenario: If a node has low available memory, the scheduler will avoid assigning a new pod to it, ensuring efficient resource utilization.

3. **Controller Manager (kube-controller-manager):** The Controller Manager runs multiple control loops (controllers) that continuously monitor the cluster state and take corrective actions when deviations occur. Each controller has a specific role in maintaining the cluster's health and desired state. The Node Controller is responsible for monitoring the health of worker nodes in a Kubernetes cluster. It continuously checks for node failures and takes corrective actions when a node becomes unresponsive. If a node fails, the controller ensures that affected pods are rescheduled to healthy nodes, maintaining the stability and availability of applications. The Replication Controller ensures that the desired number of pod replicas are always running in the cluster. If a pod crashes or is deleted unexpectedly, the controller automatically creates a replacement to match the specified configuration. This helps maintain application availability and ensures that workloads scale as needed to handle demand. The Endpoints Controller plays a crucial role in service discovery by managing the mapping between services and the pods they target. Whenever pods start, stop, or move between nodes, the controller updates endpoint records to ensure that network requests are correctly routed to active pods. This dynamic updating mechanism ensures smooth communication between services in the cluster. Example Scenario: If a pod crashes unexpectedly, the Replication Controller automatically creates a new pod to maintain the required number of instances.
4. **etcd (Cluster Store):** etcd is a highly available, distributed key-value store that acts as the central database for Kubernetes. It stores all cluster-related data, including node status, pod configurations, and network policies. Since it holds critical information, etcd must be backed up regularly to prevent data loss [32]. The etcd component is a distributed key-value store that persistently maintains the state and configuration of the Kubernetes cluster. It acts as the central database where all cluster-related information, including node statuses, resource allocations, and policy definitions, is stored. This allows Kubernetes to track and manage cluster operations efficiently. One of etcd's key roles is storing metadata, such as pod locations, service discovery details, and network configurations. Whenever changes occur, Kubernetes components retrieve and update this data to ensure that services can locate and communicate with the correct pods. This dynamic storage and retrieval mechanism enables seamless orchestration of workloads. To ensure data consistency across the cluster, etcd uses a distributed consensus protocol called Raft. This guarantees that all nodes in the cluster have a synchronized and up-to-date view of the system state. By maintaining consistency, etcd helps prevent conflicts, ensuring that Kubernetes operates reliably even in the event of node failures. Example command:

```
ETCDCTL_API=3 etcdctl get /registry/pods/default/my-pod
```

The master node components collectively ensure that Kubernetes operates efficiently by managing workloads, enforcing policies, and maintaining cluster consistency. The API Server acts as the communication hub, the Scheduler optimizes resource allocation, the Controller Manager enforces the desired state, and etcd provides persistent storage for cluster data. Together, these components enable Kubernetes to provide a scalable, self-healing, and robust orchestration environment for containerized applications.

3.3 Worker Node Components in Kubernetes

Worker nodes are responsible for executing the containers that run applications. Each worker node hosts multiple pods, which are the fundamental units of deployment in Kubernetes. To ensure efficient operation, worker nodes include three essential components: Kubelet, Kube-Proxy, and the Container Runtime. These components work together to manage pod execution, network communication, and container lifecycle [30].

1. **Kubelet:** The Kubelet is a lightweight agent running on each worker node that ensures containers are running as defined by the Kubernetes API [11]. It interacts with the API Server to receive pod definitions and manages their lifecycle by communicating with the container runtime. The Kubelet plays a crucial role in maintaining the stability of the worker node by continuously monitoring the health of running pods [33]. It ensures that all containers operate as expected, automatically restarting any failed pods to maintain application availability. This self-healing mechanism is vital for ensuring high reliability in a Kubernetes cluster. In addition to monitoring, the Kubelet is responsible for fetching and applying pod specifications received from the Kubernetes API Server. When a new pod is scheduled on a worker node, the Kubelet retrieves its configuration, including container images, resource limits, and environment variables. It then ensures that the containers are deployed according to the defined specifications. Another key function of the Kubelet is maintaining the desired state of containers as specified in Kubernetes manifests. If a container deviates from its expected state due to failures or unexpected terminations, the Kubelet intervenes to restore compliance. This ensures that applications remain consistent and operational, even in dynamic cloud-native environments. Example command:

```
systemctl status kubelet
```

2. **Kube-Proxy:** The Kube-Proxy is responsible for managing network connectivity within the Kubernetes cluster. It forwards requests to the correct pods and applies necessary network policies. By implementing iptables or other networking techniques, Kube-Proxy ensures that services remain accessible across nodes. The Kube-Proxy is a critical component of the Kubernetes worker node that manages networking and ensures seamless communication between services and pods.

One of its primary functions is forwarding network traffic within the cluster. It dynamically configures networking rules, ensuring that requests reach the appropriate pods, even as they scale or move across nodes. This helps maintain uninterrupted communication between microservices. Another essential role of Kube-Proxy is enforcing network policies to control communication between applications. By defining security rules, it restricts unauthorized access and ensures that only permitted services can interact with each other. This enhances cluster security and prevents unintended data exposure. In addition to routing and security, Kube-Proxy plays a crucial role in load balancing for service discovery. It efficiently distributes incoming requests across multiple pod replicas, ensuring that workloads are evenly distributed and preventing overloading of individual pods. This helps maintain application performance and reliability in a dynamic Kubernetes environment. Example command:

```
iptables -L -t nat
```

3. **Container Runtime:** The Container Runtime is the underlying software responsible for running containers inside pods. Kubernetes supports multiple container runtimes, including Docker, containerd, and CRI-O. The runtime pulls container images, starts containers, and manages their execution. The Container Runtime is a fundamental component of the Kubernetes worker node, responsible for executing containers within pods. One of its key functions is pulling container images from container registries such as Docker Hub, Google Container Registry, or private repositories. When a pod is scheduled, the runtime retrieves the required image, ensuring that the correct version of the application is deployed. Once the image is available, the Container Runtime launches and manages container processes within pods. It ensures that containers start with the specified configurations, including resource limits, environment variables, and networking settings. Additionally, it monitors container execution, restarting or replacing failed containers to maintain system stability. Another critical role of the Container Runtime is ensuring resource isolation and security for running containers. It enforces CPU and memory limits, preventing any single container from consuming excessive resources and affecting other workloads. The runtime also applies security policies, such as namespace isolation and process restrictions, to safeguard applications from unauthorized access and potential vulnerabilities. Example command:

```
docker ps
```

The worker node components—Kubelet, Kube-Proxy, and the Container Runtime—play a crucial role in executing and maintaining containerized applications. Kubelet ensures pod health, Kube-Proxy manages network connectivity, and the Container Runtime executes containers efficiently. Together, these components enable Kubernetes to provide a scalable and resilient platform for running modern applications.

3.4 Scalability Limits in Kubernetes

Considerations for Large Kubernetes Clusters: A cluster in Kubernetes consists of multiple nodes (physical or virtual machines) that run Kubernetes agents and are managed by the control plane [34]. Kubernetes v1.29 supports clusters with up to 5000 nodes, ensuring scalability for large-scale workloads. For optimal performance and stability, Kubernetes is designed to support clusters that meet the following conditions:

- No more than 110 pods per node—Prevents excessive resource consumption per node.
- No more than 5000 nodes—Ensures efficient orchestration and resource distribution.
- No more than 150,000 total pods—Allows high workload density within the cluster.
- No more than 300,000 total containers—Supports large-scale containerized applications.

A Kubernetes cluster can be scaled by adding or removing nodes based on deployment needs. Scaling can be done manually or automatically. Manual scaling involves provisioning or decommissioning nodes manually, which provides greater control but requires continuous monitoring and intervention. On the other hand, autoscaling uses the Cluster Autoscaler to dynamically adjust the number of nodes based on resource demand, ensuring efficient resource utilization and minimizing operational overhead. Kubernetes relies on key components to manage scaling effectively. A node is a worker machine that runs applications within pods, handling workloads and executing scheduled tasks. The control plane serves as the container orchestration layer, exposing APIs and providing an interface to define, deploy, and manage the lifecycle of containers. By leveraging these components, Kubernetes ensures seamless scalability while maintaining optimal performance and resource distribution.

3.4.1 Managing Large-Scale Kubernetes Clusters

Efficient management of large-scale Kubernetes clusters requires automation, resource optimization, and workload isolation [35]. By following best practices, organizations can improve performance, scalability, and security while minimizing operational complexity.

- **Use Cluster Autoscaler:** The Cluster Autoscaler dynamically adjusts the number of worker nodes in a cluster based on real-time resource demand. If the workload increases and existing nodes lack sufficient resources, the autoscaler automatically provisions additional nodes. Conversely, when resource utilization decreases, it removes underutilized nodes to prevent unnecessary costs and over-provisioning. This ensures efficient resource usage and cost savings, making the cluster more adaptable to workload fluctuations.

- **Distribute Workloads Across Multiple Namespaces:** Kubernetes allows users to create namespaces, which serve as logical partitions within a cluster. By distributing workloads across multiple namespaces, organizations can:
  - Improve performance by avoiding resource contention between different applications.
  - Enhance security by isolating workloads, and preventing unauthorized access or resource interference.
  - Facilitate multi-tenancy, allowing different teams or projects to operate independently within the same cluster.

  Proper namespace management ensures better resource allocation, application segmentation, and security enforcement in large-scale deployments.
- **Optimize Pod Scheduling:** The Kubernetes Scheduler assigns pods to worker nodes based on resource availability, affinity rules, and constraints to ensure efficient workload distribution. Best practices include spreading pods evenly across nodes to prevent resource bottlenecks, using affinity and anti-affinity rules to group related workloads or separate conflicting ones, and leveraging priority-based scheduling to ensure critical workloads receive resources before lower-priority tasks. By optimizing pod placement, organizations can maximize efficiency, reliability, and performance while avoiding resource congestion. By optimizing pod placement, organizations can maximize efficiency, reliability, and performance while avoiding resource congestion.

3.4.2 Example Command for Large-Scale Cluster

By adhering to these scalability guidelines, organizations can efficiently manage large Kubernetes deployments while ensuring optimal performance and resource utilization. Example: Scaling an Application

The following command scales a Kubernetes name my-app to 100 replicas:

```
kubectl scale deployment my-app --replicas=100
```

(This ensures the application can handle increased traffic by running 100 instances across the cluster. Another useful command to monitor the scaling status is:

```
kubectl get hpa
```

This command retrieves information about the Horizontal Pod Autoscaler (HPA), helping administrators track and manage automatic scaling decisions. By adhering to these scalability guidelines, organizations can efficiently manage large Kubernetes deployments while ensuring optimal performance and resource utilization.

3.5 etcd in Depth

etcd is a distributed, consistent, and highly available key-value store that serves as the backbone of Kubernetes' cluster management [32]. It ensures that cluster metadata, resource configurations, and state information are stored reliably and persistently. As Kubernetes is a dynamic system where workloads continuously change, etcd provides a reliable mechanism for storing and retrieving the latest state of the cluster.

- **State Persistence**: etcd stores all Kubernetes object states, including Pods, nodes, deployments, and services. Configurations such as ConfigMaps and Secrets and cluster-wide information such as API server settings and admission policies. Whenever changes occur in the cluster, the Kubernetes API server interacts with etcd to update and retrieve the latest state.
- **Service Discovery**: Kubernetes relies on etcd for service discovery, enabling components to retrieve information about running services and workloads. When a pod or service is scheduled, its metadata and runtime state are stored in etcd, allowing Kubernetes controllers and schedulers to fetch this data and make informed decisions [36]. This ensures seamless interaction between applications and services while maintaining a consistent and up-to-date view of the cluster.
- **Leader Election**: Kubernetes requires a control plane leader to manage scheduling, resource distribution, and failover mechanisms, and etcd plays a crucial role in leader election. It manages distributed locks to prevent multiple leaders, ensuring only one active leader at a time. In case of failure, etcd facilitates high availability and fault tolerance by electing a new leader, allowing the cluster to remain operational. Additionally, it provides consistency guarantees, ensuring that all nodes in the cluster recognize the same leader and maintain a synchronized state.

3.5.1 Consensus Algorithm: Raft

To maintain data consistency across multiple etcd instances, Kubernetes leverages the Raft consensus algorithm, a distributed consensus mechanism designed to ensure fault tolerance and high availability [37]. Raft follows a leader-follower model, where one node acts as the leader, while the remaining nodes operate as followers. This structure ensures that all changes are consistently replicated across the cluster.

- **Leader Role:** The leader is the central coordinator responsible for handling all write operations in the etcd cluster. When a client submits a request, it is directed to the leader, which processes the request, updates its own state, and then propagates the changes to all follower nodes. This approach ensures a single source of truth, preventing conflicting updates.

- **Follower Role:** Followers are passive nodes that replicate the leader's state and maintain consistency across the cluster. They do not process write requests directly but instead receive updates from the leader. If a follower falls behind, it automatically synchronizes missing data to stay up to date. Followers also play a crucial role in the leader election process by voting for a new leader when necessary.
- **Leader Election and High Availability:** If the leader node fails due to network issues or crashes, the Raft algorithm ensures that a new leader is elected automatically. The process follows these steps:
  - A follower detects the leader's absence and transitions to a candidate state.
  - The candidate solicits votes from other followers.
  - If a majority of nodes approve the candidate, it is promoted to the new leader.
  - The new leader resumes processing write requests and ensures that all nodes are synchronized.

By implementing Raft, etcd guarantees strong consistency, fault tolerance, and seamless failover. This makes Kubernetes highly resilient to failures, ensuring that critical cluster metadata remains intact even if individual nodes experience downtime.

3.5.2 Example: Storing Pod Metadata in etcd

In Kubernetes, etcd stores pod metadata as JSON objects, ensuring that the system maintains an up-to-date record of all running workloads. When a new pod is created, its metadata, including its name, namespace, and current status, is stored in etcd. This allows Kubernetes components, such as the API server and controllers, to track and manage pod states efficiently.

- **Example of Pod Metadata in etcd:** When a pod named "nginx-pod" is created in the default namespace, etcd stores its metadata in the following format:

```
{
  "metadata": {
    "name": "nginx-pod",
    "namespace": "default"
  },
  "status": "Running"
}
```

- How Kubernetes Uses This Data
  - Tracking Pod Status: The Kubernetes API server queries etcd to check whether the pod is in a Running, Pending, or Terminated state.

  - Service Discovery: Controllers and schedulers retrieve this metadata to determine where the pod is running and how it should interact with other components.
  - Fault Recovery: If a pod fails or is deleted, etcd ensures that Kubernetes has an accurate record, allowing the scheduler to restart or reschedule the workload if necessary.

  By storing structured pod metadata in etcd, Kubernetes maintains a single source of truth for managing cluster workloads, ensuring efficient orchestration and fault tolerance.

The Kubernetes API server retrieves this data to track the pod's status and location within the cluster.

3.5.3 Working with etcd: Commands and Examples

In Kubernetes, etcdctl is a command-line tool used to interact with the etcd key-value store. It allows administrators to monitor the cluster, retrieve stored data, and ensure etcd is functioning correctly. Below are some essential commands for working with etcd.

- **Checking etcd Cluster Health** To verify that the etcd service is running correctly and responding to requests, use the following command:

```
ETCDCTL_API=3 etcdctl endpoint health --endpoints=https
    ://127.0.0.1:2379
```

  Expected Output:

```
https://127.0.0.1:2379 is healthy: successfully committed
    proposal: took = 2.3ms
```

  This output confirms that the etcd instance is operational, meaning it can store and retrieve data without issues. If etcd is unhealthy, troubleshooting may involve checking logs, restarting the service, or verifying network configurations.
- **Retrieving All Keys Stored in etcd** To list all the keys currently stored in etcd, use the following command:

```
ETCDCTL_API=3 etcdctl get "" --prefix --keys-only
```

  This command retrieves all stored keys without displaying their associated values. It is useful for checking what data Kubernetes has registered in etcd, such as pod information, service configurations, and cluster states.

By using these commands, administrators can monitor etcd health, inspect stored data, and troubleshoot cluster issues, ensuring Kubernetes operates smoothly.

3.5.4 Storing and Retrieving Data in etcd

In Kubernetes, etcd acts as the central key-value store, maintaining the cluster state. Administrators can interact with etcd using the etcdctl command-line tool to store, retrieve, monitor, and delete data.

- **Storing a key-value pair in etcd:** To store data in etcd, use the put command:

```
ETCDCTL_API=3 etcdctl put /registry/pods/default/nginx-pod "
    Running"
```

  This command saves the state of the nginx-pod in the default namespace, marking it as "Running".
- **Retrieving the stored value:** To retrieve the stored value, use the get command:

```
ETCDCTL_API=3 etcdctl get /registry/pods/default/nginx-pod
```

  Expected Output:

```
/registry/pods/default/nginx-pod
Running
```

  This confirms that etcd successfully stored the pod's state.
- **Watching for Changes in etcd** To monitor real-time changes, such as pod creation, status updates, or deletion, use the watch command:

```
ETCDCTL_API=3 etcdctl watch /registry/pods/default/nginx-pod
```

  Whenever the pod's state changes, etcd, it will automatically update and display the latest value.
- **Deleting a Key from etcd** To remove a stored pod entry from etcd, use the del command:

```
ETCDCTL_API=3 etcdctl del /registry/pods/default/nginx-pod
```

  This ensures that Kubernetes no longer tracks the pod's state in etcd.

etcd plays a crucial role in Kubernetes by ensuring reliable state storage, leader election, and service discovery. By implementing the Raft consensus algorithm, etcd provides high availability and consistency, making it a fundamental component of Kubernetes architecture. Understanding how etcd stores and retrieves data is essential for effectively troubleshooting and maintaining Kubernetes clusters.

3.6 Cluster Communication

Kubernetes components interact in a structured manner to ensure smooth deployment and management of applications [38]. The communication flow within a Kubernetes cluster follows these steps:

- **User/Admin Request:** A user interacts with the Kubernetes cluster by sending commands using kubectl, API requests, or client libraries. These requests are directed to the Kubernetes API Server, which acts as the central management component of the control plane. The API Server is responsible for processing all incoming requests and coordinating with other Kubernetes components.
- **API Server Validation and Processing:** Once a request reaches the API Server, it first undergoes authentication and validation. If the request is valid, the API Server updates the etcd database, which serves as the persistent storage for cluster metadata and state. After storing the updated information, the API Server notifies the relevant Kubernetes components, such as the Scheduler or Controllers, to proceed with necessary actions.
- **Scheduler Assigns Workloads:** The Kubernetes Scheduler is responsible for assigning workloads to worker nodes. It selects the most suitable node for the pod deployment based on various factors, including CPU and memory availability, affinity rules, and resource constraints. Once a node is selected, the API Server updates the pod's state in the etcd database, allowing other components to track and manage its execution.
- **Kubelet Manages Containers:** The Kubelet is an agent running on each worker node that ensures the proper execution of assigned pods. After the API Server assigns a pod to a node, the Kubelet retrieves the pod specifications and pulls the required container image from a container registry, such as Docker Hub. Additionally, it sets up networking, storage, and other configurations needed for the containerized application to run smoothly.
- **Kube-Proxy Handles Networking:** The Kube-Proxy component plays a crucial role in managing networking within the cluster. It ensures seamless communication between pods and services by creating appropriate routing rules and handling network address translations. Additionally, Kube-Proxy enables load balancing for service traffic, ensuring efficient distribution of requests across multiple pods. These components work together to maintain an efficient, scalable, and highly available Kubernetes cluster, ensuring seamless application deployment and operation.

3.6.1 Ensuring High Availability in Kubernetes Communication

For a production-grade Kubernetes cluster, implementing high availability mechanisms is crucial to minimizing downtime and ensuring uninterrupted operations. Several key components contribute to maintaining fault tolerance and resilience in cluster communication [38].

- Multiple Master Nodes: Kubernetes clusters typically deploy multiple master nodes to ensure redundancy and prevent a single point of failure. If one master node fails due to hardware issues, network failures, or other unforeseen circumstances, another automatically takes over to continue managing the cluster. This redundancy ensures that critical control plane services, such as the API Server, Scheduler, and Controller Manager, remain operational.
- etcd Replication for Data Consistency: The etcd database is a critical component of Kubernetes, acting as a distributed key-value store that maintains cluster state and configurations. To ensure data persistence and availability, etcd operates across multiple nodes in a replicated setup. It employs the Raft consensus algorithm, which guarantees consistency by ensuring that all nodes in the cluster have an up-to-date copy of the cluster state. This mechanism prevents data loss and maintains synchronization across nodes.
- Load Balancers for Traffic Distribution: To efficiently handle user requests, load balancers are used to distribute incoming network traffic across multiple API Server instances. This ensures that requests are always directed to an available master node, preventing system overload and maintaining stable performance. By balancing traffic, load balancers help improve fault tolerance, avoid bottlenecks, and enhance the overall resilience of the cluster communication system.

By integrating these high availability mechanisms—multiple master nodes, etcd replication, and load balancing—Kubernetes ensures robust cluster communication, reducing downtime and improving reliability for enterprise-level applications.

3.6.2 Example Commands for Cluster Communication

To initiate the cluster communication process, a user can apply a deployment configuration using the following command:

- Deploying an Application: The following command applies a deployment YAML file to create a new workload:

```
kubectl apply -f deployment.yaml
```

- Checking API Server Communication: This command retrieves API Server endpoints, ensuring it is reachable:

```
kubectl get endpoints kubernetes
```

By implementing multiple master nodes, etcd replication, and network load balancing, Kubernetes achieves fault tolerance, high availability, and seamless communication between its components. In the context of high availability in cluster communication, Kubernetes architecture ensures seamless interaction between the master node and worker nodes through the API server, facilitating direct communication between end users and the cloud server. This enhances the system's availability and reliability. However, challenges may arise due to cluster communication complexities and potential single points of failure. To mitigate these risks, production Kubernetes clusters implement multiple master nodes in a leader-follower setup, ensuring redundancy and failover mechanisms. The etcd replication maintains state consistency across the cluster, preventing data loss or configuration mismatches.

3.7 Practical Example: Setting Up a Kubernetes Cluster and Troubleshooting Nodes

This practical example demonstrates how to set up a Kubernetes cluster and troubleshoot Kubernetes nodes. It provides essential steps for initializing a cluster, configuring networking [39], adding worker nodes, and verifying node status.

3.7.1 Docker Integration

Docker plays a crucial role in Kubernetes by providing container runtime support, enabling the deployment and management of containerized applications. It allows Kubernetes to run and orchestrate cloud-native applications, whether online or offline, within containers.

- Clarify Kubernetes' Shift from Docker: While Docker was historically used as the default runtime, Kubernetes deprecated direct Docker support in favor of the Container Runtime Interface (CRI), which allows it to work with other runtimes like containerd and CRI-O. You might briefly mention that Kubernetes 1.20 marked the official deprecation of Docker as a CRI runtime.
- Enhance Explanation of Pods and Containers: Kubernetes does not manage standalone containers directly; instead, it manages pods, which are the smallest deployable units containing one or more containers. A pod can also include shared storage and networking settings [39].

- Include a Kubernetes-Specific Example: While the docker run command is useful, Kubernetes does not manage containers in this way. You might replace or supplement it with:

```
kubectl run nginx-pod --image=nginx --restart=Never
```

This ensures that Kubernetes is handling the containerized workload instead of running it directly via Docker.

3.7.2 *Setting Up a Kubernetes Cluster*

Setting up a Kubernetes cluster involves configuring the control plane (master node) and adding worker nodes to form a functional environment [40]. The following steps outline the process:

- Initialize the Master Node: The master node serves as the control plane, managing the cluster and scheduling workloads. The initialization process configures essential Kubernetes components, including the API Server, Scheduler, and Controller Manager.

```
kubeadm init --pod-network-cidr=192.168.1.0/24
```

  This command initializes the cluster, setting up the control plane and defining the pod network CIDR.
- Set Up Networking: Kubernetes requires a network plugin to enable pod-to-pod communication across nodes. Deploying Calico, a commonly used network policy, ensures seamless connectivity:

```
kubectl apply -f https://docs.projectcalico.org/manifests/
    calico.yaml
```

  This applies the Calico network configuration, enabling inter-pod communication within the cluster.
- Join Worker Nodes: To expand the cluster, worker nodes must be registered with the master node. Each worker node uses a unique token and hash for authentication:

```
kubeadm join <master-ip>:6443 --token <token> --discovery-
    token-ca-cert-hash <hash>
```

  This command securely connects the worker nodes to the master, allowing them to execute scheduled workloads.

- Verify Cluster Nodes: Once the worker nodes are added, their status can be checked using the following command:

```
kubectl get nodes
```

  This lists all nodes in the cluster and displays their readiness status, ensuring they are properly configured.

By following these steps, a fully functional Kubernetes cluster can be set up, enabling efficient container orchestration and workload distribution.

3.7.3 Troubleshooting Kubernetes Nodes

Effectively diagnosing and resolving node-related issues in a Kubernetes cluster requires a structured approach. The following are common debugging techniques:

- Check Node Status: To inspect the current state of all nodes, including their IP addresses and health conditions, use:

```
kubectl get nodes -o wide
```

  This command provides detailed information on each node's status, including readiness, roles, and assigned IP addresses.
- Describe a Pod for Errors: If a pod is failing or not running as expected, obtain detailed insights into events, errors, and resource allocations with:

```
kubectl describe pod <pod-name>
```

  This command helps identify scheduling issues, failed container startups, and insufficient resource allocations.
- View Cluster Logs: To check real-time logs of a running container inside a pod, use:

```
kubectl logs -f <pod-name>
```

  This command streams live logs from the container, helping diagnose runtime errors and application-specific issues

Setting up a Kubernetes cluster involves initializing the master node, configuring networking, and adding worker nodes, while effective troubleshooting focuses on checking node status, describing pods for errors, and analyzing logs to diagnose and resolve issues efficiently.

3.8 Key Takeaways

- Kubernetes Architecture: A deep dive into the master and worker node components, including the API server, scheduler, controller manager, etcd, kubelet, container runtime, and kube-proxy, highlighting their roles in cluster orchestration.
- Cluster State Management: Understanding etcd as a distributed key-value store, which ensures data consistency using the Raft consensus algorithm and facilitates efficient cluster operations.
- Networking and Communication: Explanation of cluster communication mechanisms, including Kube-Proxy for managing network traffic and Calico networking for secure pod-to-pod communication.
- Docker Integration: Insights into container runtime support, showcasing how Kubernetes manages containerized applications and supports alternatives like containerd and CRI-O.
- Practical Implementation: Step-by-step guidance on setting up a Kubernetes cluster, initializing the master node, configuring networking, and adding worker nodes.
- Debugging and Troubleshooting: Essential techniques for checking node status, describing pods for errors, and viewing cluster logs, ensuring smooth Kubernetes operations.
- Scalability and Optimization: Strategies for efficient workload distribution, pod scheduling, and autoscaling, including best practices for large-scale Kubernetes deployments.

3.9 Summary

This chapter provides a comprehensive analysis of Kubernetes' core components, beginning with the master node, which orchestrates cluster operations, schedules workloads, and enforces policies, and the worker node, which executes containerized applications while maintaining connectivity with the control plane. A key focus is on etcd, detailing its role as Kubernetes' central data store and its implementation of the Raft consensus algorithm to ensure high availability and consistency. The chapter delves into cluster communication mechanisms, explaining how nodes interact efficiently to sustain system integrity. It further explores container runtime integration, highlighting Kubernetes' support for Docker, containerd, and CRI-O in managing containerized applications. A hands-on section guides readers through setting up a Kubernetes cluster, covering network configuration, pod scheduling, and scaling strategies. Additionally, the chapter introduces troubleshooting methodologies, providing practical approaches for diagnosing node failures and resolving common cluster issues. Finally, it examines scalability challenges, presenting best practices for autoscaling, resource optimization, and workload distribution to enhance Kubernetes' performance in large-scale environments.

3.10 Review Questions

- What are the four core components of the Kubernetes control plane and their functions?
- Explain how the kubelet on worker nodes interacts with the API server on the master node.
- What is the role of container runtime on worker nodes and which runtimes are supported?
- What factors typically limit scalability in Kubernetes clusters?
- How does Kubernetes handle node failures in large-scale deployments?
- What command would you use to check the current resource utilization across 1000+ nodes?
- Why is etcd critical to Kubernetes operation and what happens if it fails?
- Explain how the Raft consensus algorithm ensures data consistency in etcd.
- What etcdctl command would you use to retrieve metadata about a specific pod?
- How does Kubernetes implement leader election for control plane components?
- What kubectl command shows communication paths between cluster components?
- How does Kubernetes ensure secure communication between nodes and control plane?
- What are the minimum system requirements for setting up a Kubernetes cluster with Docker?
- What command initializes a Kubernetes control plane node and what flags are crucial for production?
- How would you troubleshoot a worker node that shows "NotReady" status?

Chapter 4
Kubernetes vs. Docker Swarm

Abstract This chapter provides a comparative study of Kubernetes and Docker Swarm, the two most widely used container orchestration platforms. It begins with an overview of both systems and outlines their roles in managing containers alongside Docker. A structured comparison highlights differences in installation complexity, including step-by-step commands for each platform, as well as GUI support and ecosystem maturity. Scalability and performance are discussed in depth, showing how Kubernetes excels in large-scale environments while Docker Swarm offers simplicity for smaller deployments. Load balancing and service discovery mechanisms are compared, followed by an analysis of storage and data management approaches in both platforms. The chapter also emphasizes monitoring and observability, explaining how each system handles cluster visibility and operational insights. By addressing features, challenges, and trends, this chapter equips readers with a clear understanding of the strengths and trade-offs between Kubernetes and Docker Swarm in real-world scenarios.

Keywords Kubernetes · Docker Swarm · Container orchestration · Load balancing · Installation complexity · GUI support · Monitoring

This chapter provides an overview of Kubernetes and Docker Swarm, including their installation setup, GUI support, scalability, and performance. Additionally, it explores load balancing [41], storage management, and monitoring capabilities in both orchestration tools. It details how Kubernetes and Docker Swarm handle persistent storage through volume management, stateful applications, and data replication strategies. Furthermore, it explains observability features such as logging, metrics collection, and monitoring tools that help maintain system health and optimize performance.

B. Kumar et al., *Modern Kubernetes: From Core Concepts to Intelligent Autoscaling for Cloud Applications*, Studies in Autonomic, Data-driven and Industrial Computing, https://doi.org/10.1007/978-3-032-12972-7_4

4.1 Overview of Kubernetes and Docker Swarm

Container orchestration plays a crucial role in modern software deployment, allowing organizations to efficiently manage large-scale containerized applications. While containers offer portability and scalability, manual management becomes challenging as deployments grow. Orchestration automates essential tasks such as deployment, scaling, networking, load balancing, and monitoring. Among the leading orchestration tools, Kubernetes and Docker Swarm stand out, each offering unique capabilities. Kubernetes, originally developed by Google and now maintained by the Cloud Native Computing Foundation (CNCF), has emerged as the industry standard due to its self-healing, auto-scaling, rolling updates, and advanced networking features. With widespread adoption—used by approximately 90% of organizations for containerized workloads—Kubernetes is a powerful solution for enterprises managing complex applications. However, its steep learning curve and installation complexity require a solid understanding of its architecture and functionalities. Its key strengths include auto-scaling, fault tolerance, extensive networking support, flexibility across environments, strong community backing from major tech companies, and the ability to manage thousands of nodes efficiently. Docker Swarm is Docker's native container orchestration tool, designed for simplicity and ease of use. Unlike Kubernetes, which has a complex architecture, Docker Swarm offers a lightweight solution, making it well-suited for small-scale deployments [42]. Swarm mode enables users to cluster multiple Docker hosts, allowing containerized applications to scale horizontally with minimal effort. One of its key advantages is seamless integration with Docker, as it utilizes native Docker commands and tools, ensuring a familiar workflow for developers. Additionally, its simple setup, requiring just a single command (docker swarm init), makes it highly accessible for teams looking for a quick and efficient orchestration solution [43]. However, despite these advantages, Docker Swarm has a smaller adoption rate, with only 10% market share compared to Kubernetes. The primary drawback of Docker Swarm is its lack of advanced features, such as built-in monitoring, auto-healing, and complex networking support, which limits its suitability for enterprise-grade applications requiring higher scalability and resilience.

4.1.1 Comparison for Docker, Docker Swarm and Kubernetes

Docker is best suited for running individual containers, making it ideal for local development, testing, and small applications due to its lightweight and user-friendly nature, though it lacks built-in orchestration for multi-host environments. **Docker Swarm**, Docker's native clustering tool, provides simple orchestration for small to medium-scale deployments with easy setup and management, but it falls short in scalability and advanced automation needed for large-scale production. In contrast, **Kubernetes (K8s)** is a powerful yet complex orchestration platform designed for

Table 4.1 Comparison for Docker, Docker Swarm and Kubernetes

Feature	Docker	Docker Swarm	Kubernetes (K8s)
Installation time	Minimal setup	Single command	Lengthy, multi-step
Configuration	Basic setup	Minimal configuration	Complex (RBAC, networking)
Ease of Use	Beginner-friendly	Simple	Requires expertise
Purpose	Container runtime for running single containers	Native clustering for Docker containers (orchestration)	Enterprise-grade container orchestration
Scalability	Limited to single host	Scales to multiple nodes (simpler scaling)	Highly scalable (handles large clusters efficiently)
Complexity	Simple and lightweight	Easy setup, minimal configuration	Complex setup but highly configurable
Load balancing	Manual or basic networking	Automatic internal load balancing	Advanced load balancing (Ingress, Services)
High availability	Not built-in (relies on external tools)	Basic HA with manager/worker nodes	Robust HA (self-healing, auto-replacement)
Rolling updates	Manual updates per container	Supports rolling updates	Advanced rolling updates (canary, blue-green)
Networking	Basic bridge/host networking	Overlay networks (multi-host)	Advanced CNI (supports multiple plugins)
Storage	Local or basic volume mounts	Supports volumes across nodes	Dynamic storage provisioning (PV/PVC)
Monitoring	Basic logs/stats	Limited monitoring	Extensive monitoring (Prometheus, Grafana)
Use case	Local development, single apps	Small to medium clusters (simple setups)	Large-scale, production-grade deployments

large-scale, production-grade deployments, offering advanced features like auto-scaling, self-healing, and rolling updates, making it the industry standard despite its steeper learning curve. The choice between these tools depends on the scale and complexity of your deployment needs (See Table 4.1).

4.1.2 Latest Trends and Ecosystem Updates in Docker and Kubernetes

This section discusses the latest trends and updates in the Docker and Kubernetes ecosystem. This topic covers Docker's lightweight containerization, Kubernetes' dynamic resource reallocation, and the Kubernetes Gateway. Additionally, we compare security aspects across different versions of Docker and Kubernetes while exploring edge-supported Kubernetes for AI/ML workloads.

1. Docker and Kubernetes Features:

   - Docker's Shift to Containers: Since Docker Engine v20.10, Docker has adopted containers as its default runtime, replacing its legacy components. This shift aligns Docker more closely with Kubernetes, which has always relied on contained as its core runtime. By leveraging containers, Docker benefits from improved performance, with lower overhead and faster container startup times. It also enhances compatibility, allowing seamless integration with Kubernetes clusters due to the shared runtime. Additionally, security is strengthened, as Containerd's minimalistic design reduces the attack surface. This change simplifies workflows for developers using Docker for local development and Kubernetes for production, ensuring a more streamlined container ecosystem.
   - Kubernetes Features: This feature is discussed in the coming chapter in detail.

     – Sidecar Containers (Stable in Kubernetes v1.28+): Kubernetes v1.28 introduces stable sidecar containers, enhancing pod architecture by enabling auxiliary containers to run alongside primary ones. These sidecars share network and storage, supporting tasks like logging (Fluentd), monitoring (Prometheus), and traffic management (Envoy, Istio). Security tools like Falco can also run as sidecars for threat detection. This approach reduces resource overhead and simplifies operations, benefiting service mesh and observability pipelines. A key update is native lifecycle management, ensuring sidecars start before and stop after primary containers, eliminating manual ordering and improving reliability.
     – Dynamic Resource Allocation (v1.26+): Stable since v1.26, dynamic resource allocation enhances GPU, FPGA, and network interface management, enabling secure, fine-grained resource sharing. It improves GPU utilization beyond the "one GPU per pod" model, with a pluggable driver framework for hardware vendors (NVIDIA, Intel, AMD). Kubernetes' scheduling framework now supports dynamic GPU allocation, workload preemption, hot-swapping, and QoS guarantees via ResourceClaim and ResourceClass APIs, forming an enterprise-grade orchestration layer for AI/ML workloads.
     – Kubernetes Gateway API (Replacing Ingress): Kubernetes Gateway API (Replacing Ingress): Generally available in v1.28, the Gateway API replaces Ingress, offering greater flexibility and control over networking. It separates infrastructure concerns from application development, enabling cluster admins to define policies while developers focus on routing. Key features include weighted routing, header-based matching, and cross-namespace traffic control. Supporting HTTP, gRPC, and TCP/UDP, it

surpasses Ingress limitations and integrates with Istio, NGINX, and other controllers, solidifying its role as Kubernetes' future networking standard.

- Wasm (WebAssembly) Support: Both Docker and Kubernetes now support WebAssembly (Wasm) workloads, enabling lightweight, sandboxed applications with enhanced security and efficiency. Docker integrates Wasm through containerd's Wasm runtimes like wasmtime and wasmedge, allowing developers to run Wasm containers alongside Linux containers while maintaining compatibility with existing workflows. Kubernetes supports Wasm through projects like Kwasm and SpinKube, which use Wasm-specific runtimes (e.g., containerd-shim-spin) and custom schedulers for high-density deployments. Wasm is gaining traction in edge computing and serverless functions due to its fast startup times (<1 ms) and strong security isolation, making it an efficient alternative to traditional containers for high-performance.

2. Security Comparison: Docker and Kubernetes have enhanced security features to protect containerized workloads. Docker introduced Rootless mode (v20.10+), reducing privileges for better isolation, and Slim/Scout CLI for vulnerability scanning. Kubernetes strengthens security with Pod Security Admission (PSA), replacing PodSecurityPolicy (PSP) for stricter access control, along with Network Policies powered by Cilium and Calico to enforce fine-grained traffic control. Additionally, SIG Security updates ensure continuous improvements and vulnerability fixes, making both platforms more secure for modern cloud-native applications.
3. Edge Computing Support Kubernetes: Several lightweight Kubernetes distributions and container orchestration tools cater to edge computing and resource-constrained environments. K3s (by SUSE), MicroK8s (by Canonical), and k0s (by Mirantis) offer streamlined Kubernetes deployments with reduced overhead, making them ideal for IoT and edge scenarios. In contrast, Docker Swarm has limited edge capabilities but includes Docker Edge Stack for basic edge management, providing a simpler alternative for container orchestration at the edge.
4. AI/ML Workload Support Kubernetes: Kubernetes enhances machine learning (ML) workflows with powerful integrations. Kubeflow simplifies ML deployment by providing a native framework for training and serving models. The NVIDIA GPU Operator optimizes GPU management, ensuring seamless resource allocation for AI workloads. Additionally, auto-scaling for ML pipelines dynamically adjusts resources based on demand, improving efficiency and performance in cloud-native AI applications.
5. Serverless and FaaS: Kubernetes supports serverless computing through frameworks like Knative, OpenFaaS, and Kubeless, enabling Function-as-a-Service (FaaS) workloads. Docker, while offering OpenFaaS integration, has limited native serverless support, requiring additional tools for full functionality. Knative extends Kubernetes with event-driven autoscaling, while OpenFaaS simplifies function deployment, making it easier to run serverless workloads across different environments.

6. Multi-Cluster Management: Kubernetes excels in multi-cluster management with tools like Cluster API, Karmada for multi-cluster orchestration, and Anthos/GKE Fleet (Google) for hybrid deployments. In contrast, Docker Swarm lacks native multi-cluster management, relying on third-party tools for such capabilities. These Kubernetes solutions provide automated cluster scaling, workload distribution, and cross-cluster networking, making them ideal for large-scale, multi-cloud environments

The latest trends and ecosystem updates in Docker and Kubernetes highlight significant advancements in resource management, security, serverless computing, multi-cluster orchestration, and AI/ML integration. Kubernetes continues to evolve with dynamic resource allocation, sidecar container stabilization, and WebAssembly (Wasm) support, improving performance and efficiency. Enhanced security mechanisms such as Pod Security Admission and network policies further strengthen containerized environments. Meanwhile, serverless frameworks like Knative and OpenFaaS are driving Kubernetes' dominance in cloud-native computing. Docker, while maintaining its role in container development, has aligned more closely with Kubernetes by adopting containerd as its runtime and integrating Wasm workloads. However, its multi-cluster management and native serverless capabilities remain limited compared to Kubernetes. As cloud-native architectures continue to advance, Kubernetes remains the industry leader in scalability, automation, and workload orchestration, while Docker continues to refine its developer-centric tools. The ongoing innovations in both ecosystems will further shape the future of edge computing, AI workloads, and enterprise cloud strategies.

4.2 Installation Complexity and Commands

The installation process of **Kubernetes** and **Docker Swarm** differs significantly, with Kubernetes requiring multiple components and configurations, while Docker Swarm offers a simpler setup [44].

4.2.1 Kubernetes Installation

Setting up Kubernetes requires several steps because it consists of multiple components working together, including setting up key components such as `kubeadm`, `kubelet`, and `kubectl`. The major commands used for installation are:

- **Initialize the cluster:** The first step is to set up the control plane node, which acts as the "brain" of the Kubernetes cluster. This node manages all worker nodes and schedules workloads. To initialize the control plane, run:

```
kubeadm init
```

This command does several things: Sets up cluster certificates and security settings. Configures networking for communication between nodes and Creates essential Kubernetes components like the API server and scheduler. After running this command, Kubernetes provides instructions to configure kubectl (the command-line tool) so that you can manage the cluster. This command sets up the control plane node, generating certificates, networking, and cluster configurations.

- **Set up networking:** Kubernetes needs a Container Network Interface (CNI) to allow communication between pods running on different nodes. A networking plugin like Calico, Flannel, or Weave Net must be installed. To apply a network plugin, use:

```
kubectl apply -f <network-plugin.yaml>
```

This ensures that all pods can communicate properly, preventing issues where some services cannot reach others.

- **Join worker nodes to the cluster:** Once the control plane is ready, additional nodes (worker nodes) can be added to the cluster. Worker nodes are responsible for running applications. To add a worker node, run the command provided by Kubernetes when initializing the cluster:

```
kubeadm join <master-node-ip>:6443 --token <token> --
    discovery-token-ca-cert-hash sha256:<hash>
```

This connects the worker node to the Kubernetes control plane, allowing it to receive and run workloads.

- **Verify Installation:** Once all nodes are added, check the cluster status with:

```
kubectl get nodes
```

If everything is set up correctly, you should see the control plane and worker nodes listed as “Ready”.

Since Kubernetes setup can be complex, many companies prefer using managed Kubernetes services, such as: Google Kubernetes Engine (GKE)—Kubernetes hosted on Google Cloud. Amazon Elastic Kubernetes Service (EKS)—Kubernetes managed by AWS. Azure Kubernetes Service (AKS)—Microsoft Azure’s Kubernetes solution. These services handle much of the setup and maintenance, making it easier to deploy and manage Kubernetes clusters without worrying about manual configurations.

- Why is Kubernetes Installation More Complex than Docker Swarm?
  - Multiple components: Kubernetes requires setting up a control plane, worker nodes, and networking, whereas Docker Swarm can be initialized with a single command.
  - Networking configuration: Unlike Swarm, Kubernetes needs an external networking plugin to manage pod communication.
  - Security and authentication: Kubernetes requires certificates and role-based access control (RBAC), making it more secure but also more complex.

4.2.2 Docker Swarm Installation

Unlike Kubernetes, Docker Swarm is built into Docker itself, making it much easier to set up and manage. Swarm allows you to deploy and scale containerized applications with minimal configuration.

- **Initialize a Swarm Cluster:** To create a Swarm cluster, you need a manager node, which controls the cluster and schedules services. Run the following command on the node that will act as the manager:

```
docker swarm init
```

  This command: Sets up the manager node. Generates a unique token for worker nodes to join the cluster. Configures networking for inter-node communication. Once the manager node is initialized, Docker provides a command with a join token, which is required to connect worker nodes.
- **Add Worker Nodes to the Swarm:** Worker nodes are responsible for running application containers. To add a new worker node to the Swarm cluster, run:

```
docker swarm join --token <token> <manager-node-ip>:2377
```

  This command: Connects the worker node to the manager node. Allows the worker node to receive and execute tasks assigned by the manager. This command allows worker nodes to join the cluster with minimal configuration.
- **Verify the Swarm Cluster:** Once all nodes are added, check the cluster status by running this command on the manager node:

```
docker node ls
```

  This will list all nodes in the Swarm, showing whether they are managers or workers and their current status.

- Why is Docker Swarm Easier to Install than Kubernetes?

  - Single Command Setup—Unlike Kubernetes, which requires multiple components and configurations, Docker Swarm is enabled with just docker swarm init.
  - No External Networking Plugins—Swarm has built-in networking, eliminating the need for third-party network configurations.
  - Lightweight and Faster Deployment—Since Swarm is integrated into Docker, it requires fewer resources and is easier to deploy compared to Kubernetes.

- When to Use Docker Swarm Over Kubernetes?
  - For quick and simple container orchestration—Swarm is best for small to medium-scale deployments where ease of use is a priority.
  - For Docker-native environments—If you are already using Docker, Swarm integrates seamlessly with Docker Compose and other tools.
  - For lightweight deployments—Swarm is less resource-intensive than Kubernetes, making it ideal for environments with limited resources.

4.2.3 Comparison of Installation Complexity

The installation complexity of Kubernetes and Docker Swarm differs significantly in terms of setup time, configuration requirements, and ease of use [44]. The table below summarizes these differences:

```
Feature            Kubernetes                  Docker Swarm
Installation Time  Lengthy, multiple steps     Quick, single command
Configuration      Complex (RBAC, networking) Minimal setup
Ease of Use        Requires expertise          Beginner-friendly
```

While Docker Swarm is easy to install and configure, Kubernetes offers advanced capabilities such as:

- Auto-scaling—Automatically adjusts resources based on demand.
- Self-healing—Restarts failed containers and ensures application stability.
- Advanced Networking—Supports multiple networking models and policies.

Organizations must decide between Docker Swarm and Kubernetes based on their scalability, complexity, and operational requirements.

4.3 GUI Support in Kubernetes and Docker Swarm

Graphical User Interface (GUI) support plays a crucial role in container orchestration by providing a visual representation of workloads, system resources, deployments, and overall cluster health. Instead of relying solely on command-line interfaces (CLI), a GUI helps users interact with containerized environments in a more intuitive and user-friendly manner.

4.3.1 Kubernetes GUI Support

Kubernetes includes a built-in Dashboard UI, which allows users to monitor and manage cluster resources efficiently. It provides an overview of applications running in the cluster and displays detailed information about workloads, services, and resource utilization.

- Installing the Kubernetes Dashboard [22]: To deploy the Kubernetes Dashboard, users must apply the official Dashboard YAML file:

```
kubectl apply -f dashboard.yaml
```

  After installation, the dashboard can be accessed through a web browser, but authentication and security configurations must be set up before use.
- Accessing the Dashboard Securely: Since the Kubernetes Dashboard does not have public access by default, users need to create a secure access method: Create an authentication token for login

```
kubectl -n kubernetes-dashboard create token
    admin-user
```

  This generates a token that users can copy and use to log in to the dashboard.
- Start a Proxy to Securely Access the Dashboard

```
kubectl proxy
```

  This command starts a proxy server that enables access to the dashboard through the local machine.
- Access the Dashboard via a Web Browser: Open the following link in a browser:

```
http://localhost:8001/api/v1/namespaces/kubernetesdashboard/
    services/https:kubernetes-dashboard:/proxy/
```

- Role-Based Access Control (RBAC) Setup [46, 47]: By default, Kubernetes restricts dashboard access using RBAC policies. Users must create proper roles and permissions to securely manage the cluster through the dashboard.
- Features of the Kubernetes Dashboard
  - Cluster Overview: Displays running applications, resource usage, and event logs.
  - Workload Management: Enables users to scale deployments, restart pods, and update configurations.

– Service Monitoring: Provides insights into network services, persistent storage, and logs.
– RBAC and Security [48]: Ensures restricted access to specific users based on assigned roles.

4.3.2 Docker Swarm GUI Support

Unlike Kubernetes, Docker Swarm does not provide a built-in GUI for cluster management. Instead, users rely on third-party tools to visualize and manage their Swarm environments.

- Using Portainer for Docker Swarm [49]: Portainer is one of the most popular GUI tools for Docker and Docker Swarm. It provides an easy-to-use web interface for managing containers, services, networks, and volumes.
- Installing Portainer on Docker Swarm: To deploy Portainer, users need to run the following command:

```
docker volume create portainer_data
docker run -d -p 9000:9000 --name=portainer --restart=always
   \
   -v /var/run/docker.sock:/var/run/docker.sock \
   -v portainer_data:/data portainer/portainer-ce
```

  Once installed, Portainer can be accessed via a web browser at:

```
            http://localhost:9000
```

- Features of Portainer for Docker Swarm
  – Node Management: Provides a visual overview of Swarm nodes and services.
  – Container Control: Allows users to create, start, stop, and restart containers easily.
  – Network and Volume Management: Simplifies the process of configuring networks and persistent storage.
  – Stack Deployment: Enables users to deploy services using Docker Compose.

This built-in interface provides an intuitive way to oversee deployments, services, and overall cluster health. However, additional authentication and role-based access control (RBAC) configurations are required to enable secure access. On the other hand, Docker Swarm does not include an official GUI by default. Instead, users rely on third-party tools such as Portainer, which provides a web-based interface for managing Docker containers and Swarm clusters. Portainer simplifies container monitoring, deployment, and scaling operations, making it a popular choice for Docker Swarm users. Overall, Kubernetes offers native GUI support through its dashboard, while Docker Swarm depends on external tools for visualization and management.

4.4 Scalability and Performance

Scalability is a critical factor in container orchestration, ensuring that an application can efficiently handle increasing workloads. Both Kubernetes and Docker Swarm support horizontal scaling, which involves adding more container instances to meet demand. However, their scalability capabilities and approaches differ significantly [34]. Kubernetes is designed for large-scale, production-grade workloads and can support clusters with over 5000 nodes. Its architecture enables efficient workload distribution and dynamic scaling based on resource utilization. Some of its key scalability features include:

- Horizontal Pod Autoscaler (HPA) [50, 51]: This mechanism automatically adjusts the number of running pods based on CPU utilization or custom-defined metrics, ensuring optimal resource usage.
- Cluster Autoscaler [52]: Unlike HPA, which adjusts pods within nodes, the Cluster Autoscaler dynamically scales the number of worker nodes in response to workload demands.
- Advanced Scheduling Algorithms: Kubernetes uses sophisticated scheduling policies to distribute workloads across available nodes efficiently, reducing the likelihood of performance bottlenecks.

These features make Kubernetes a robust solution for applications requiring high availability, fault tolerance, and seamless scaling in cloud-native environments. Docker Swarm, though more lightweight in design, supports clusters of up to 2000 nodes. It provides a faster container scaling mechanism but lacks Kubernetes' advanced autoscaling capabilities. Some key scalability features of Docker Swarm include:

- Swarm Mode: This built-in feature allows users to scale containers quickly using simple Docker commands, making it an intuitive solution for developers who prioritize ease of use.
- Service Replication: Docker Swarm automatically replicates container instances to balance workloads efficiently across the cluster, ensuring service availability.

While Docker Swarm offers faster scaling speeds per container, it lacks dynamic node scaling and intelligent workload distribution features found in Kubernetes. This makes Docker Swarm more suitable for small to medium-scale applications where simplicity and rapid scaling are preferred over complex orchestration.

4.5 Load Balancing

Load balancing is a crucial mechanism in container orchestration, ensuring that incoming requests are evenly distributed across available resources. Efficient load balancing prevents performance bottlenecks, optimizes resource utilization, and enhances application availability. While both Kubernetes and Docker Swarm provide load-balancing mechanisms, their approaches and implementation differ significantly [21].

4.5.1 Load Balancing in Kubernetes

Load balancing is a crucial function in Kubernetes to ensure efficient distribution of network traffic across multiple pods and nodes. It helps in optimizing resource usage, improving fault tolerance, and maintaining high availability of applications. Kubernetes provides multiple service types to manage traffic flow based on the specific needs of the application.Kubernetes offers four main service types for distributing traffic:

- ClusterIP: The default service type, which distributes traffic internally within the Kubernetes cluster. This is useful for communication between different microservices.

```
apiVersion: v1
kind: Service
metadata:
  name: my-internal-service
spec:
  selector:
    app: my-app
  ports:
    - protocol: TCP
      port: 80
      targetPort: 8080
  type: ClusterIP
```

  Use Case: When you have backend services (e.g., databases, APIs) that should only be accessible within the cluster.
- NodePort: Exposes a service on a static port across all cluster nodes, allowing external access without needing a cloud provider's load balancer.

```
apiVersion: v1
kind: Service
metadata:
  name: my-nodeport-service
spec:
  selector:
    app: my-app
  ports:
    - protocol: TCP
      port: 80
      targetPort: 8080
      nodePort: 30000
  type: NodePort
```

Accessing the Service: http://<node-ip>:30000 Use Case: When you want to test services externally without setting up a cloud load balancer and deploying in an on-premise environment where cloud-based load balancers are not available.

- LoadBalancer: Integrates with cloud provider services (such as AWS ELB, Google Cloud Load Balancer, or Azure Load Balancer) to route external traffic efficiently.

```yaml
apiVersion: v1
kind: Service
metadata:
  name: my-loadbalancer-service
spec:
  selector:
    app: my-app
  ports:
    - protocol: TCP
      port: 80
      targetPort: 8080
  type: LoadBalancer
```

Use Case: When deploying web applications, APIs, and microservices that require public access with automatic traffic distribution and Ideal for cloud-based environments where external traffic needs to be efficiently routed.

- Ingress Controller: Provides advanced traffic management, including SSL termination, request routing, and URL-based access control. This is essential for managing microservices that require flexible traffic routing and security policies.

```yaml
apiVersion: networking.k8s.io/v1
kind: Ingress
metadata:
  name: my-ingress
  annotations:
    nginx.ingress.kubernetes.io/rewrite-target: /
spec:
  rules:
  - host: myapp.example.com
    http:
      paths:
      - path: /
        pathType: Prefix
        backend:
          service:
            name: my-service
            port:
              number: 80
  tls:
  - hosts:
    - myapp.example.com
    secretName: my-tls-secret
```

Use Case: When running microservices that require domain-based routing (e.g., api.example.com, shop.example.com) and When handling HTTPS traffic with SSL termination for enhanced security. Ideal for enterprises that need fine-grained traffic control.

Kubernetes provides a powerful and flexible load-balancing system, but manual configurations are often needed, especially for Ingress controllers and custom routing setups. Organizations must carefully choose their load balancing method based on their scalability needs, security policies, and cloud infrastructure.

4.5.2 Load Balancing in Docker Swarm

Docker Swarm provides a simple yet effective load balancing mechanism that ensures even distribution of traffic across multiple service replicas in a cluster. Unlike Kubernetes, which offers multiple load balancing options, Docker Swarm uses an inbuilt Routing Mesh that automatically handles traffic forwarding and service discovery.

- Routing Mesh: Ensures that requests sent to any node in the cluster are automatically forwarded to an available container instance, regardless of the node where it is running [54].

```
    docker service create --name my-service --replicas 3
        -p 8080:80 my-app
```

  How It Works: The service is created with 3 replicas, each running on different nodes. Incoming traffic to port 8080 on any node will be forwarded to one of the replicas. Use Case: Ideal for high-traffic applications that need automatic load distribution.
- Service Discovery: Automatically assigns virtual IPs to services and maintains a dynamic list of healthy containers for traffic routing.

```
version: "3"
services:
  web:
    image: nginx
    ports:
      - "8080:80"
    deploy:
      replicas: 3
  backend:
    image: my-backend
    networks:
      - my-network
networks:
  my-network:
```

How It Works: The web service can automatically discover the backend service by its name (backend). Swarm manages an internal DNS to route traffic efficiently without needing manual IP configurations.
Use Case: Useful for microservices architectures, where services must dynamically discover each other.

- Simplified Configuration: Unlike Kubernetes, Docker Swarm requires minimal setup for load balancing, making it ideal for environments where ease of deployment is a priority.

```
        docker service scale my-service=5
```

This command increases the number of running replicas to 5. Swarm automatically balances traffic among the new instances. Use Case: Ideal for environments that prioritize ease of deployment and automation.

While Docker Swarm's automatic load balancing is easier to implement, it lacks the fine-grained control and customization options available in Kubernetes.

```
Feature                Kubernetes                Docker Swarm
Traffic Routing    manual configuration    handled by Routing Mesh
External Load Balancing cloud-based Load Balancers Limited support
Advanced Traffic Control Ingress for complex rules Basic load balancing
Ease of Setup          Setup \& outer integrations Need minimal configure
```

While Docker Swarm's built-in Routing Mesh simplifies traffic distribution, Kubernetes provides more control and flexibility, making it better suited for complex architectures requiring advanced traffic routing and security policies.

4.6 Storage and Data Management in Kubernetes and Docker Swarm

Persistent storage is essential for **stateful applications**, ensuring data remains available even when containers restart or migrate across nodes. Kubernetes and Docker Swarm handle storage differently, with Kubernetes offering a more **robust and flexible** solution compared to Docker Swarm's simpler volume-based approach.

4.6.1 Storage in Kubernetes

Kubernetes provides a flexible and scalable storage system to ensure that containerized applications can access and persist data, even if pods are restarted or moved across nodes. Unlike traditional storage, which is tightly coupled to a specific machine, Kubernetes decouples storage from compute resources, making it highly portable across cloud and on-premises environments [55].

- **Persistent Volumes (PVs):** A Persistent Volume (PV) is a cluster-wide storage resource that exists independently of the lifecycle of any pod [55]. Ensures that data remains available even when pods are restarted, rescheduled, or moved to a different node [56]. Defined by administrators and supports various storage backends like NFS, iSCSI, AWS EBS, Google Persistent Disk, and CephFS.

```
apiVersion: v1
kind: PersistentVolume
metadata:
  name: example-pv
spec:
  capacity:
    storage: 10Gi
  accessModes:
    - ReadWriteOnce
  persistentVolumeReclaimPolicy: Retain
  storageClassName: standard
  hostPath:
    path: "/mnt/data"
```

  How It Works: This PV provides 10GB of storage and is retained even after the pod using it is deleted. It is accessible in ReadWriteOnce mode, meaning only one node can write to it at a time.
- **Persistent Volume Claims (PVCs):** Allows applications to request storage dynamically.
- **Dynamic Storage Provisioning:** A Persistent Volume Claim (PVC) is a request for storage by a pod. Allows applications to dynamically request and use storage without needing to know the underlying storage type [56]. Kubernetes automatically binds a PVC to an available PV with matching storage specifications.

```
apiVersion: v1
kind: PersistentVolumeClaim
metadata:
  name: example-pvc
spec:
  accessModes:
    - ReadWriteOnce
  resources:
    requests:
      storage: 5Gi
  storageClassName: standard
```

  How It Works: The PVC requests 5GB of storage with ReadWriteOnce access mode. Kubernetes automatically finds a matching PV (if available) and binds it.
- **Dynamic Storage Provisioning**—Automating storage allocation, Kubernetes can automatically provision storage on demand using Storage Classes. Eliminates the need for pre-creating Persistent Volumes, making storage more scalable and it Works with cloud providers (AWS, GCP, Azure) and on-premises solutions (Ceph, GlusterFS, NFS).

```
apiVersion: storage.k8s.io/v1
kind: StorageClass
metadata:
  name: fast-storage
provisioner: kubernetes.io/aws-ebs
parameters:
  type: gp2
  fsType: ext4
```

How It Works: Defines a Storage Class called fast-storage that uses AWS Elastic Block Store (EBS) and When a PVC requests storage using fast-storage, Kubernetes automatically provisions the storage. Use Case: Automatically creating database volumes, persistent caches, or backup storage in cloud environments.

Kubernetes provides a powerful and flexible storage system that allows applications to persist data efficiently. By leveraging Persistent Volumes, Persistent Volume Claims, and Dynamic Storage Provisioning, organizations can ensure scalable, resilient, and cloud-compatible storage solutions for containerized workloads.

4.6.2 Storage in Docker Swarm

Docker Swarm provides a simplified storage model for containerized applications, focusing on ease of use and performance. However, compared to Kubernetes' flexible storage system, Docker Swarm's approach has limitations, particularly in dynamic provisioning and portability.

- **Global Volumes:** Docker Swarm uses volumes to persist data, but these volumes are tied to specific nodes. Ensures data remains available to a container even after it restarts. Since volumes are node-bound, if a container moves to another node, it may lose access to its original storage. Example: Creating a Volume in Docker Swarm

```
        docker volume create --name my_volume
```

  This creates a named volume (my_volume) that can be attached to a container. Use Case: Storing application data, logs, or database files that must persist beyond container lifecycles.
- **No Dynamic Provisioning:** To use a volume in a Docker Swarm service, it must be explicitly mounted when deploying a container. Example: Using a Volume in a Swarm Service

```
docker service create --name my_service \
--mount type=volume,source=my_volume,target=/app/data \
  nginx
```

How It Works: my_service runs an Nginx container with a persistent volume (my_volume). The volume is mounted to /app/data, ensuring that files stored there persist. Use Case: Ensuring web applications or databases retain their configuration and data.

```
Feature          Kubernetes                        Docker Swarm
Storage Type     Pod-level Persistent Volumes      Node-bound Global
    Volumes
Persistence      High (Persistent Volumes          Lower (Data is tied to
    nodes)
                  allow data retention)
Dynamic Provision Yes (Cloud & on-prem support) No (Requires manual
    setup)
```

Kubernetes offers a more advanced and scalable storage solution, making it well-suited for stateful applications that require high availability and persistent data storage. Its use of Persistent Volumes (PVs) and Persistent Volume Claims (PVCs) allows for dynamic storage provisioning, ensuring seamless data management across cloud and on-premises environments. In contrast, Docker Swarm relies on global volumes, which, while simpler and easier to set up, are node-dependent and lack flexibility. This makes Swarm more suitable for smaller deployments that do not require complex storage management.

4.7 Monitoring and Observability

Monitoring plays a crucial role in tracking resource utilization, system health, and application performance. Kubernetes has built-in monitoring capabilities, making it more efficient for large-scale deployments, while Docker Swarm relies on third-party solutions for observability [57]. Kubernetes integrates seamlessly with monitoring tools such as Prometheus and Grafana. Prometheus collects time-series metrics from nodes, pods, and services, while Grafana provides real-time visualization dashboards. Additionally, Kubernetes supports log and metrics collection through Fluentd, the ELK (Elasticsearch, Logstash, Kibana) stack, and OpenTelemetry, ensuring comprehensive system observability. These built-in features allow Kubernetes users to monitor workloads efficiently without requiring extensive external configurations. In contrast, Docker Swarm does not offer native monitoring tools, requiring users to rely on third-party solutions such as the ELK Stack, Datadog, and cAdvisor. While these tools provide valuable insights into system performance, their integration demands additional configuration, making monitoring more complex compared to Kubernetes. Consequently, Kubernetes is better suited for organizations that require detailed observability, automated alerting, and seamless performance tracking, whereas Docker Swarm users must invest additional effort in setting up and maintaining monitoring solutions.

Use Case Scenarios Choosing between Kubernetes and Docker Swarm depends on the scale, complexity, and operational requirements of an application. Docker Swarm is well-suited for small to medium-sized deployments where simplicity, quick setup, and ease of management are priorities. Its lightweight architecture allows for fast container orchestration with minimal overhead, making it an excellent choice for development and testing environments. Teams that require a straightforward solution without extensive configuration may find Docker Swarm more efficient. On the other hand, Kubernetes is the preferred option for large-scale, enterprise-grade applications that demand advanced features such as auto-scaling, high availability, and security. It is designed to handle complex workloads across hybrid or multi-cloud environments while providing robust management of resources, networking, and persistent storage. Organizations that require fine-grained control over deployment strategies and need to optimize resource utilization for production workloads will benefit from Kubernetes' flexibility and scalability.

4.8 Key Takeaways

- Comparative Analysis of Kubernetes and Docker Swarm: The chapter examines key differences between the two container orchestration platforms in terms of installation, scalability, load balancing, storage, and monitoring.
- Installation Complexity: Kubernetes requires multiple components and configurations, making installation more complex, while Docker Swarm offers a simpler setup with fewer commands.
- GUI Support: Kubernetes provides multiple dashboard options, whereas Docker Swarm has limited built-in graphical interfaces.
- Scalability and Performance: Kubernetes is designed for large-scale, production-grade workloads, while Docker Swarm is better suited for smaller, less complex environments.
- Load Balancing: Kubernetes utilizes Services and Ingress controllers for load balancing, whereas Docker Swarm relies on an internal routing mesh.
- Storage Management: Kubernetes offers advanced storage solutions, including persistent volumes and dynamic provisioning, while Docker Swarm provides simpler but less flexible storage options.
- Monitoring and Observability: Kubernetes integrates with mature monitoring tools like Prometheus and Grafana, whereas Docker Swarm has fewer built-in monitoring capabilities.

4.9 Summary

This chapter compares Kubernetes and Docker Swarm as container orchestration platforms, highlighting their strengths and weaknesses. It begins with installation complexity, explaining that Kubernetes requires a more intricate setup compared to Docker Swarm's lightweight deployment. The discussion on GUI support shows that Kubernetes has better visualization tools, while Docker Swarm's interface is minimal. The chapter then covers scalability and performance, emphasizing Kubernetes' robustness in large-scale distributed applications versus Docker Swarm's efficiency in simpler deployments. Load balancing mechanisms differ significantly, with Kubernetes offering external and internal balancing through Services, while Docker Swarm provides automatic internal load balancing via routing mesh. Storage and data management are more advanced in Kubernetes, supporting persistent storage, dynamic volume claims, and cloud integration, while Docker Swarm's storage options are more basic. The final section on monitoring and observability highlights Kubernetes' strong ecosystem with tools like Prometheus and Grafana, whereas Docker Swarm has limited built-in monitoring capabilities. Overall, the chapter provides a clear understanding of when to use Kubernetes versus Docker Swarm based on project requirements.

4.10 Review Questions

- What are the key architectural differences between Kubernetes and Docker Swarm?
- Compare the cluster formation process in Kubernetes vs. Docker Swarm.
- Why is Kubernetes generally considered more complex to install than Docker Swarm?
- What are the minimum system requirements for setting up a Kubernetes cluster versus a Docker Swarm cluster?
- How does the command-line interface (CLI) experience differ between kubectl and docker swarm commands?
- What GUI tools are available for Kubernetes cluster management?
- Compare the built-in dashboard capabilities of Kubernetes and Docker Swarm.
- How does third-party tool integration differ between Kubernetes and Docker Swarm for GUI management?
- At what scale does Kubernetes typically outperform Docker Swarm in terms of performance?
- What makes Kubernetes better suited for large-scale, production-grade deployments?
- Compare the auto-scaling capabilities of Kubernetes and Docker Swarm.
- How does load balancing implementation differ between Kubernetes Services and Docker Swarm?
- What advanced load balancing features does Kubernetes offer that Docker Swarm lacks?
- Compare the persistent storage solutions available in Kubernetes versus Docker Swarm.
- What monitoring and observability tools integrate natively with Kubernetes but not with Docker Swarm?

Chapter 5
Kubernetes Version Evolution and Comparative Analysis

Abstract This chapter traces the evolution of Kubernetes versions and highlights the impact of major feature milestones across its lifecycle. It begins with foundational releases (v1.14–v1.20), focusing on early stability improvements, persistent volume advancements, and networking API maturity. Intermediate versions (v1.21–v1.25) are analyzed for the graduation of key features and the deprecation of Docker in favor of containerd. Advanced releases (v1.26–v1.30) introduce dynamic resource allocation for GPUs and FPGAs, the stabilization of sidecar containers, and the progression from Ingress to the Gateway API. A timeline of security improvements, along with performance benchmarking, illustrates the platform's resilience and efficiency gains. The chapter also examines upgrade paths, ensuring smooth transitions across versions, while ecosystem alignment is discussed through CSI, CNI, CRI dependencies, Helm compatibility, and cloud provider timelines. Finally, the roadmap for v1.31 and beyond is presented, outlining planned features and SIG Architecture directions.

Keywords Container runtime · Containerd · Persistent volumes · Networking APIs · Dynamic resource allocation · Sidecar containers · Gateway API · Security evolution · CSI · CNI · CRI · Helm · Operator SDK

Kubernetes has undergone significant evolution since its inception, with each version bringing improvements in stability, scalability, security, and performance. This analysis provides a detailed breakdown of Kubernetes' versioning strategy, major milestones, and future roadmap, helping organizations make informed decisions on upgrades and compatibility. This chapter provides a comprehensive analysis of Kubernetes version evolution, starting with an introduction to versioning. It examines foundational versions (v1.14–v1.20), highlighting early stability, persistent volume improvements, and networking API maturity. Intermediate versions (v1.21–v1.25) are analyzed for major feature stabilization and the transition from Docker runtime to containerd. Advanced capabilities (v1.26–v1.30) cover dynamic resource allocation, sidecar container stabilization, security advancements, and performance

B. Kumar et al., *Modern Kubernetes: From Core Concepts to Intelligent Autoscaling for Cloud Applications*, Studies in Autonomic, Data-driven and Industrial Computing, https://doi.org/10.1007/978-3-032-12972-7_5

benchmarking. Additionally, it explores ecosystem alignment, including CSI, CNI, CRI dependencies, and cloud provider support. Finally, it outlines the future roadmap (v1.31+), discussing planned features and SIG architecture directions.

5.1 Introduction to Kubernetes Versioning

- Release Cadence and Support Lifecycle: Kubernetes follows a structured release process, ensuring continuous innovation and stability [58]. Typically, Kubernetes releases four minor versions per year, each introducing new features, enhancements, and deprecations.
  - Minor Versions (e.g., v1.28 → v1.29): These versions bring new capabilities, API enhancements, and improvements to cluster management.
  - Patch Versions (e.g., v1.28.3): Focused on bug fixes, performance optimizations, and security patches, ensuring system reliability.
  - Long-Term Support (LTS): While Kubernetes does not have an official LTS model, cloud providers like AWS (EKS), Google Cloud (GKE), and Azure (AKS) extend support beyond the default lifecycle.

  Each version is supported for approximately 14 months, with Kubernetes maintaining updates for only the three most recent minor releases. This means organizations must plan regular upgrades to stay within the supported range and receive security updates.
- Significance of Version Comparisons: Understanding Kubernetes version differences is critical for operational stability and security:
  - Feature Enhancements: Organizations can leverage new capabilities such as improved scheduling, networking APIs, and security controls.
  - Deprecations and Breaking Changes: Comparing versions helps teams avoid deprecated features and outdated APIs, preventing disruptions.
  - Security and Compliance: Upgrading ensures compliance with CVE patches, RBAC improvements, and policy enforcement mechanisms.
  - Performance Optimization: Each release brings refinements in resource utilization, cluster efficiency, and networking stack enhancements.

  By staying informed about Kubernetes versioning, organizations can make strategic decisions about upgrades, workload migration, and ecosystem compatibility.

5.2 Foundational Version Comparison (v1.14–v1.20)

Between Kubernetes v1.14 and v1.20, the platform underwent significant enhancements to improve stability, particularly in storage, networking, and workload orchestration [59]. Persistent storage support was strengthened through refined Persistent Volume Claims (PVCs) and the introduction of Volume Snapshot APIs, enabling better data management, resilience, and backup strategies. Networking improvements, such as CoreDNS becoming the default cluster DNS provider and advancements in Kubernetes Ingress, enhanced service discovery, load balancing, and multi-cluster traffic routing. Additionally, workload orchestration saw optimizations in scheduling algorithms, reducing pod startup latency and improving resource allocation, making Kubernetes more efficient and production-ready.

5.2.1 Key Milestones in Early Stability

Between Kubernetes v1.14 and v1.20, significant improvements were made to stabilize critical functionalities for production workloads.

- Container Storage Interface (CSI) Support: A key milestone was the introduction of the Container Storage Interface (CSI), enabling seamless integration with third-party storage providers while ensuring vendor-agnostic storage, simplified driver deployment, and dynamic provisioning. Enhancements in StatefulSets and Persistent Volumes (PVs) improved data consistency, expanded storage class features, and refined lifecycle management for stateful workloads. Additionally, workload scheduling mechanisms were optimized to reduce pod startup latency, enhance resource utilization through efficient bin-packing, and implement preemption logic, allowing high-priority workloads to execute seamlessly.
- Improvements in Stateful Applications and Persistent Volumes: Kubernetes progressively improved the management of StatefulSets and Persistent Volumes (PVs) to enhance data consistency, reliability, and storage flexibility. These enhancements ensured stable storage for stateful applications, expanded storage class features to support dynamic volume provisioning, and refined lifecycle management, allowing better control over volume creation, retention, and deletion. These advancements made Kubernetes more robust for handling persistent workloads in production environments.
- Enhanced Workload Scheduling Mechanisms: Kubernetes enhanced its scheduling algorithms to improve efficiency and responsiveness in workload management. These optimizations reduced pod startup latency, ensuring faster application deployment, implemented more efficient bin-packing to maximize resource utilization, and introduced preemption logic, allowing high-priority workloads to execute when resources became available. These improvements contributed to better performance, scalability, and overall cluster efficiency.

5.2.2 Persistent Volume Evolution

Storage advancements in Kubernetes were a key focus area from v1.14 to v1.20 [60].

- PVC Refinements (v1.14–v1.16): Between Kubernetes v1.14 and v1.16, Persistent Volume Claims (PVCs) were refined to enhance storage management and flexibility. These improvements introduced dynamic storage provisioning, enabling automatic volume creation based on demand, enhanced storage reclaim policies for better control over persistent data handling, and strengthened volume binding modes, ensuring more efficient scheduling decisions for pods utilizing Persistent Volumes (PVs). These advancements significantly improved storage reliability and efficiency in Kubernetes environments.
- Volume Snapshot API (v1.20): With the release of Kubernetes v1.20, the Volume Snapshot API was introduced, providing a robust mechanism for data backup and recovery. This feature enabled point-in-time volume snapshots, allowing users to restore data without disrupting running applications, making it essential for disaster recovery strategies. Additionally, the API supported vendor-specific snapshot implementations, offering greater flexibility for cloud-based storage solutions and improving overall data resilience in Kubernetes environments

5.2.3 Networking API Maturity

The evolution of networking APIs played a vital role in improving Kubernetes' overall stability [61]. Enhancements in network policies and communication mechanisms strengthened system reliability.

- CoreDNS as the Default DNS Provider: Kubernetes officially transitioned CoreDNS to a stable release, replacing kube-dns as the default DNS provider. This upgrade brought better performance through improved caching and optimized query resolution, enhanced flexibility with custom plugins for DNS-based service discovery, and easier extensibility, enabling users to configure DNS behavior without modifying Kubernetes internals. These improvements streamlined service discovery and boosted overall cluster networking efficiency.
- Ingress and Service Networking Enhancements: Kubernetes introduced significant Ingress and service networking enhancements to improve external traffic routing and cluster connectivity. Load balancing optimizations enabled more efficient multi-backend load balancing strategies, enhancing traffic distribution. Multi-cluster traffic routing improvements allowed seamless communication across multiple Kubernetes clusters, increasing scalability and flexibility. Additionally, improved TLS handling introduced better SSL certificate management, strengthening security for HTTP traffic. These advancements made Kubernetes networking more robust, scalable, and secure.

Between Kubernetes v1.14 and v1.20, the platform made significant strides toward stability, especially in storage, networking, and workload management. These improvements laid the foundation for modern Kubernetes deployments, making it a reliable choice for enterprises looking to run stateful applications, optimize networking, and ensure high availability.

5.3 Intermediate Feature Analysis (v1.21–v1.25)

During Kubernetes versions v1.21 to v1.25, the platform matured significantly by stabilizing critical features and optimizing core components for production environments. This period focused on refining workload management, enhancing security, improving extensibility, and making Kubernetes more efficient and reliable.

5.3.1 Graduation of Major Features to Stable

Several previously experimental features were officially promoted to stable releases, ensuring their reliability and long-term support.

- Ephemeral Containers (v1.23): These containers provided an essential tool for on-the-fly debugging of running pods. Instead of restarting a pod to inject debugging tools, users could dynamically add ephemeral containers to troubleshoot live applications, significantly enhancing Kubernetes' observability and maintenance capabilities.
- CronJobs Stabilization (v1.21): Kubernetes' CronJobs, used for scheduling periodic tasks, became stable after multiple refinements. This feature improved the scheduling of batch jobs, ensuring better job execution reliability, resource efficiency, and failure handling in production environments.
- Immutable ConfigMaps and Secrets (v1.21): To enhance security and configuration management, Kubernetes introduced immutable ConfigMaps and Secrets. This feature prevented accidental or malicious modifications of critical configurations, ensuring that sensitive data, such as database credentials and API keys, remained tamper-proof throughout their lifecycle.

5.3.2 Deprecation of Docker Runtime (containerd Transition)

One of the most impactful changes in Kubernetes v1.24 was the removal of native Docker runtime support in favor of containerd and CRI-O. This transition was driven by the need for a more lightweight, efficient, and Kubernetes-native runtime that adhered to the Container Runtime Interface (CRI) standard [62].

- Why Kubernetes moved away from Docker? Docker, while widely used, introduced unnecessary overhead, such as a built-in daemon, that Kubernetes did not require. In contrast, containerd and CRI-O were optimized for Kubernetes, offering faster startup times, reduced resource consumption, and improved container lifecycle management. The shift to containerd aligned Kubernetes with a more modular and runtime-agnostic ecosystem, enhancing long-term maintainability. Although this transition required administrators to update their

container runtime configurations, it ultimately resulted in a more efficient, scalable, and Kubernetes-native orchestration system.
- Custom Resource Definitions (CRDs) became first-class citizens in the Kubernetes API, replacing the need for API aggregation, which previously introduced complexity and performance overhead. This shift allowed developers to create custom Kubernetes objects without modifying the core API, streamlining extensibility. Additionally, Kubernetes introduced validation schemas for CRDs, enforcing structural integrity and reducing misconfigurations, ultimately enhancing cluster reliability and ensuring consistency in custom resource definitions.

The v1.21–v1.25 era marked Kubernetes' evolution into a more mature, secure, and extensible platform. With ephemeral containers, CronJobs, and immutable configurations reaching stability, Kubernetes improved its reliability in production. The deprecation of Docker in favor of containerd streamlined container runtime management, reducing unnecessary overhead. Finally, the advancements in CRDs empowered users to extend Kubernetes more efficiently, reinforcing its flexibility as a cloud-native orchestration system. These changes collectively made Kubernetes faster, more secure, and better suited for enterprise-scale deployments.

5.4 Advanced Capabilities (v1.26–v1.30)

Kubernetes versions v1.26 to v1.30 introduced significant advancements aimed at improving resource management, networking, and security. These enhancements make Kubernetes more efficient, scalable, and capable of handling modern workloads, especially for AI/ML applications, microservices, and cloud-native architectures.

5.4.1 Dynamic Resource Allocation (GPUs/FPGAs)

As workloads become more complex, hardware acceleration has become essential for optimizing computational performance, particularly for machine learning (ML) and artificial intelligence (AI) applications [63]. Kubernetes has improved support for specialized hardware resources, including GPUs, TPUs, and FPGAs, enabling better workload distribution and cost-efficient resource utilization.

- GPU and FPGA Support: Kubernetes now allows better integration of hardware accelerators (e.g., NVIDIA GPUs and Intel FPGAs) for AI and high-performance computing workloads.
- NVIDIA GPU Operator: Streamlines GPU provisioning and management, reducing the need for manual configurations.

- Improved Autoscaling: Kubernetes enhances autoscaling mechanisms to dynamically adjust GPU/FPGA resources based on workload demand, ensuring optimal performance and cost efficiency.
  With these advancements, Kubernetes can efficiently manage AI-driven applications, reducing infrastructure overhead and improving model training speeds.

5.4.2 Sidecar Container Stabilization

Starting from v1.28, Kubernetes natively supports sidecar containers, improving the service mesh, observability, and security workloads [64].

- What Are Sidecars? A sidecar container is an auxiliary container deployed alongside an application container in the same pod. It extends the functionality of the main container without modifying its core codebase. Sidecars are commonly used for: Logging and monitoring (e.g., Prometheus, Fluentd), Service mesh (e.g., Istio, Linkerd) and Security and policy enforcement (e.g., Envoy).
- Why Is Native Support Important? Previously, sidecars were managed at the application level, leading to inconsistent behavior because: Sidecars and main containers started and stopped independently, sometimes causing failures and Lifecycle management was not tightly integrated with Kubernetes' control mechanisms.
- Key Improvements in v1.28+
  - Synchronized Lifecycle Management: Sidecars now start before and terminate after the primary container, ensuring a smooth shutdown process and avoiding unexpected failures.
  - Better Integration with Service Meshes: Enhanced traffic routing and observability for microservices architectures.
  - Improved Security: Enables policy enforcement and zero-trust networking through integrated security proxies.

These enhancements simplify sidecar deployment and management, making service meshes and monitoring stacks more reliable in Kubernetes environments.

5.4.3 Gateway API vs. Ingress Progression

Kubernetes networking architecture has evolved with the introduction of the Gateway API, which became generally available (GA) in v1.28. It is designed to replace the traditional Ingress API, offering greater flexibility and scalability for managing network traffic [65].

- Limitations of Ingress API:
  - Limited traffic routing options (mainly HTTP-based).
  - Lack of fine-grained access control (RBAC was not built-in).
  - Namespace constraints (Ingress is tied to a single namespace).
- Advantages of Gateway API:
  - Advanced Traffic Routing: Supports layer-7 routing, TCP/UDP, and mTLS encryption for secure service-to-service communication.
  - Role-Based Access Control (RBAC): Offers granular security controls, allowing different teams to manage network policies without affecting the entire cluster.
  - Multi-Namespace Support: Unlike Ingress, which operates within a single namespace, Gateway API enables multi-namespace configurations, improving network isolation network and scalability.
- Adoption by Major Tools:
  - Istio, NGINX, and Contour have already adopted Gateway API, signaling its role as the future of Kubernetes networking.
  - Cloud providers (AWS, Google Cloud, Azure) are integrating Gateway API into their managed Kubernetes services.
  - The Gateway API is expected to fully replace Ingress in future Kubernetes releases, providing enterprise-grade networking capabilities.

The advanced capabilities in Kubernetes v1.26–v1.30 enhance resource management, security, and networking, enabling efficient AI/ML workload execution, improving service mesh stability with native sidecar support, and replacing Ingress with Gateway API for better security, scalability, and traffic management, making Kubernetes more efficient and adaptable for modern cloud infrastructure.

5.5 Security Evolution Timeline

Security Evolution in Kubernetes Security has been a key focus in recent Kubernetes updates, with improvements in pod security, network isolation, and vulnerability management to enhance cluster resilience [66].

- Pod Security Policy $\rightarrow$ Pod Security Admission Kubernetes deprecated Pod Security Policies (PSP) in v1.21 and introduced Pod Security Admission (PSA) in v1.25. PSA simplifies security enforcement by validating pod security settings at deployment, eliminating the need for complex cluster-wide PSP configurations. It also integrates seamlessly with Role-Based Access Control (RBAC), making policy enforcement more efficient and manageable.
- Network Policy Enhancements Advanced security features in networking have been introduced, leveraging eBPF-powered tools like Cilium and Calico for

fine-grained network segmentation and real-time traffic filtering. Additionally, Kubernetes has strengthened service mesh security by enforcing mutual TLS (mTLS), ensuring encrypted communication between microservices, mitigating unauthorized access, and preventing lateral movement attacks within clusters.
- CVE Resolution and Vulnerability Management Kubernetes actively addresses Common Vulnerabilities and Exposures (CVEs) through frequent patch updates and version-specific security advisories.

This approach ensures clusters remain protected against emerging threats, offering a clear security roadmap for administrators to follow while maintaining compliance with security best practices.

5.6 Performance Benchmarking

Kubernetes has continuously optimized resource efficiency, scheduling latency, and cluster scalability, ensuring better performance for large-scale deployments [67].

- Scheduling Speed Improvements: Kubernetes enhanced pod scheduling speed by introducing Parallel Scheduling (v1.20+), allowing multiple pods to be scheduled simultaneously, significantly reducing decision-making time in large clusters. Additionally, node-local caching mechanisms improved latency by enabling schedulers to make faster, more informed placement decisions without querying the entire cluster state repeatedly.
- Resource Utilization Efficiency: Recent Kubernetes versions have optimized CPU and memory request handling, minimizing resource wastage and ensuring better allocation across workloads. These improvements help maintain high efficiency, particularly in dynamic environments where workloads frequently scale in and out. Advanced autoscaling policies for cost-efficient infrastructure management.
- Cluster Density Optimization Trends: Kubernetes has continually refined its resource management strategies to enhance cluster scalability and reduce API server load, enabling higher node densities per cluster. Efficient API server load management has been achieved through API request optimization, which reduces unnecessary API calls and leverages improved caching mechanisms to lower the control plane load. Additionally, watch-based event propagation minimizes API server stress by relying on watch-based updates instead of frequent polling. Advanced bin-packing techniques further optimize pod placement, with topology-aware scheduling ensuring workloads are evenly distributed across available resources to minimize fragmentation. Enhancements like resource-aware scheduling (v1.23+) improve CPU and memory allocation, leading to better node utilization, while NUMA-aware scheduling (v1.26+) enhances performance by placing workloads closer to their required hardware resources. These advancements collectively improve Kubernetes' ability to support large-scale clusters efficiently.

By implementing these cluster density optimizations, Kubernetes supports larger clusters with better efficiency, reducing infrastructure overhead while maintaining high availability and performance.

5.7 Upgrade Path Analysis: Ensuring Smooth Kubernetes Transitions

Upgrading Kubernetes is a critical process that requires careful planning to prevent service disruptions caused by breaking changes, deprecated features, and new security enforcements. Kubernetes follows a release cycle where new versions introduce enhancements while occasionally removing outdated functionalities [68].

- Breaking Changes by Major Version: Several key Kubernetes releases have introduced significant changes that require adaptation:
  - v1.24—Dockershim Removal: Kubernetes deprecated Dockershim, the built-in Docker runtime, requiring administrators to migrate to containerd or CRI-O. This change improved container runtime efficiency but necessitated updating runtime configurations.
  - v1.25—PodSecurityPolicy (PSP) Deprecation: Kubernetes removed Pod Security Policies (PSP) in favor of Pod Security Admission (PSA), requiring cluster operators to transition to the new policy framework to enforce workload security.
  - v1.27+—Stricter RBAC Policies: Role-Based Access Control (RBAC) policies were tightened by default, reducing potential security risks but necessitating review and adjustment of existing permissions for applications and users.
- Recommended Upgrade Sequences: To minimize risks during Kubernetes upgrades, administrators should follow best practices, including stepwise upgrades, avoiding skipping multiple versions to prevent compatibility issues (e.g., upgrading directly from v1.22 to v1.26). Before rolling out updates in production, testing in a staging environment is essential to identify potential issues with workloads and configurations. Additionally, taking etcd backups and preparing rollback strategies ensures quick recovery in case of unexpected failures, reducing downtime and maintaining cluster stability.
- Version Compatibility Matrices: Each cloud provider, such as Amazon EKS, Google Kubernetes Engine (GKE), and Azure Kubernetes Service (AKS), maintains version compatibility matrices outlining supported Kubernetes versions and dependencies. These matrices help organizations plan upgrades by ensuring compatibility with managed Kubernetes services and infrastructure components like networking policies, storage drivers, and monitoring tools.

By following structured upgrade strategies and consulting compatibility guidelines, organizations can seamlessly transition between Kubernetes versions while maintaining security, stability, and performance.

5.8 Ecosystem Alignment

As Kubernetes evolves, its broader ecosystem—including storage, networking, and orchestration tools—must stay aligned to ensure compatibility and performance. Several key interfaces and tools rely on version-specific updates to function optimally within Kubernetes clusters.

5.8.1 CSI, CNI, and CRI Interface Version Dependencies

CSI, CNI, and CRI interface updates enhance storage portability, networking performance, and container runtime efficiency in Kubernetes [69].

- Container Storage Interface (CSI): Updates to the CSI improve storage portability by standardizing how storage providers integrate with Kubernetes. This ensures that persistent storage volumes can be dynamically provisioned across different cloud and on-premises environments without custom drivers.
- Container Networking Interface (CNI): Networking plugins like Cilium and Calico leverage eBPF technology to enhance security, observability, and performance. Kubernetes versions continuously refine CNI integration, enabling better network isolation, multi-cluster communication, and service mesh compatibility.
- Container Runtime Interface (CRI): Kubernetes deprecating Dockershim in favor of containerd and CRI-O (v1.24+) ensures better runtime efficiency and modularity. This shift reduces runtime overhead, making container orchestration more lightweight and scalable.

5.8.2 Helm and Kubernetes Operator SDK Compatibility

Helm and Kubernetes Operator SDK require version-specific updates to ensure seamless integration and compatibility with Kubernetes releases [70].

- Helm Chart Versioning: Helm, Kubernetes' package manager, requires version-specific updates to keep charts compatible with evolving API changes (e.g., deprecations of old Kubernetes API versions).
- Operator SDK Adaptation: Kubernetes Operators, which automate complex application management tasks, depend on specific Kubernetes APIs. With each

new Kubernetes release, Operators must update their logic to support changes in API resources and security policies.

5.8.3 Cloud Provider Support Timelines

Cloud providers like AWS EKS, Google GKE, and Azure AKS follow Kubernetes' version deprecation policies, ensuring managed services remain up-to-date and secure [71].

- Managed Kubernetes Services: Cloud providers such as AWS EKS, Google GKE, and Azure AKS follow Kubernetes' official deprecation and support timelines. These platforms enforce end-of-life (EOL) policies, requiring users to upgrade clusters periodically.
- Upgrade Windows: Cloud providers often provide grace periods before removing support for older Kubernetes versions, ensuring organizations have sufficient time to test and upgrade workloads without disruption.

By maintaining compatibility across storage, networking, orchestration, and cloud providers, Kubernetes ensures a stable and scalable ecosystem that supports modern cloud-native applications.

5.9 Future Roadmap (v1.31+)

As Kubernetes continues to evolve, future releases will prioritize AI/ML workloads, high-performance networking, and policy-driven security. The Special Interest Groups (SIGs) are actively shaping the roadmap to enhance scalability, interoperability, and efficiency across clusters [72].

5.9.1 Planned Features in Alpha/Beta

- GPU-Aware Scheduling Enhancements: Kubernetes will introduce fine-grained GPU resource allocation to optimize ML/AI workloads, improving scheduling efficiency for GPUs, TPUs, and FPGAs.
- Improved WebAssembly (Wasm) Support: WebAssembly will become a first-class citizen in Kubernetes, allowing developers to run lightweight, secure workloads with faster cold-start times and lower memory consumption.

5.9.2 SIG Architecture Direction

SIG Architecture focuses on enhancing multi-cluster federation, improving scalability, and refining Kubernetes' core design for future workloads.

- Multi-Cluster Federation Improvements: Kubernetes will enhance multi-cluster federation, enabling organizations to seamlessly manage workloads across multiple Kubernetes clusters with improved networking, observability, and policy enforcement.
- Policy-Driven Security Models: Future releases will refine Pod Security Admission and RBAC policies, ensuring stricter security postures while reducing operational complexity.

With these advancements, Kubernetes aims to streamline AI/ML infrastructure, enhance multi-cluster management, and strengthen security, ensuring it remains the leading container orchestration platform for cloud-native applications.

5.10 Key Takeaways

- Persistent Volume and Storage Enhancements: Kubernetes improved PVC handling, Volume Snapshots, and reclaim policies, enhancing data persistence and recovery.
- Networking and Service Evolution: CoreDNS replaced kube-dns for better performance and extensibility, while Ingress and Gateway API advancements optimized traffic routing and security.
- Runtime and CRD Improvements: The Docker runtime was deprecated in favor of containerd and CRI-O, and Custom Resource Definitions (CRDs) became first-class citizens, improving Kubernetes extensibility.
- Security Enhancements: The transition from PodSecurityPolicy (PSP) to Pod Security Admission (PSA) simplified security enforcement, while Cilium and Calico introduced eBPF-powered network security.
- Performance Optimization: Kubernetes reduced API server load, improved scheduling latency, and enhanced cluster density with advanced bin-packing algorithms.
- Upgrade Path Considerations: Major versions introduced breaking changes (e.g., Dockershim removal in v1.24, PSP deprecation in v1.25), requiring careful upgrade sequencing and staging tests.
- Ecosystem Alignment: Updates to CSI (storage), CNI (networking), and CRI (container runtime) improved integration with cloud-native tools, ensuring seamless compatibility.
- Future Roadmap (v1.31+): Kubernetes is set to enhance AI/ML workload scheduling, support WebAssembly (Wasm), and improve multi-cluster federation and security policies.

5.11 Summary

This chapter explored the evolution of Kubernetes across multiple versions, highlighting key advancements in storage, networking, security, performance, and ecosystem integration. Kubernetes enhanced persistent storage with Volume Snapshots and CSI improvements, while networking saw upgrades like the adoption of CoreDNS, advanced Ingress handling, and Gateway API support. The deprecation of Docker runtime in favor of containerd and CRI-O streamlined container orchestration, and Custom Resource Definitions (CRDs) became integral to Kubernetes extensibility. Security evolved with the transition from PodSecurityPolicy (PSP) to Pod Security Admission (PSA) and enhanced network isolation via Cilium and Calico. Performance optimizations included better API load management, scheduling enhancements, and improved cluster density. The chapter also discussed upgrade best practices, emphasizing the need for stepwise upgrades and staging environment testing to mitigate risks. Additionally, the alignment of Kubernetes with cloud providers through CSI, CNI, and CRI advancements ensures smooth ecosystem integration. Looking ahead, future releases (v1.31+) will focus on AI workload optimization, WebAssembly (Wasm) support, and multi-cluster federation tools, reinforcing Kubernetes' role as a scalable and future-ready container orchestration platform.

5.12 Reviews Questions

- What is the significance of Volume Snapshot API (v1.20) in Kubernetes?
- How did CoreDNS replace kube-dns, and what are its key advantages?
- What improvements were made to Ingress and service networking in recent Kubernetes versions?
- How did Kubernetes transition from Docker runtime to containerd and CRI-O, and why?
- What are Custom Resource Definitions (CRDs), and how have they evolved in Kubernetes?
- How did Kubernetes enhance Pod Security by transitioning from PodSecurityPolicy (PSP) to Pod Security Admission (PSA)?
- What network security enhancements were introduced with tools like Cilium and Calico?
- How did Kubernetes optimize API server load management to support higher cluster densities?
- What role does NUMA-aware scheduling (v1.26+) play in Kubernetes performance improvements?
- What are the recommended best practices for upgrading Kubernetes clusters?
- How do CSI, CNI, and CRI advancements align Kubernetes with cloud providers like AWS, GCP, and Azure?

- What are the benefits of sidecar container stabilization (v1.28+) for service mesh architectures?
- How does the Gateway API improve upon the traditional Ingress API in Kubernetes networking?
- What future roadmap features (v1.31+) are planned for AI workloads and multi-cluster management?
- How does Kubernetes' evolving ecosystem impact Helm charts and Operator SDK compatibility?

Chapter 6
Effective Stateful Applications and Data Persistence

Abstract This chapter focuses on building and managing stateful applications in Kubernetes with an emphasis on data persistence and lifecycle management. It begins by explaining container filesystem layers, differentiating between system and application containers, and demonstrating Kubernetes' role in orchestrating real-world web applications. The use of kubectl is detailed, covering namespaces, contexts, object management, and advanced filtering with JSONPath to efficiently organize environments. Pods are examined as the core execution units, including resource allocation, lifecycle control, and multi-container designs with shared storage. Labels, annotations, and selectors are introduced as powerful tools for metadata management and grouping objects. Service discovery methods, ReplicaSets for scaling, DaemonSets for distributed workloads, and Jobs for batch processing are presented with practical use cases. The chapter also addresses configuration management through ConfigMaps and secrets, while deployments are shown as the foundation for managing updates. Finally, persistent data challenges are explored, along with strategies to ensure reliable storage, highlighting Kubernetes' ability to balance scalability with durability.

Keywords Stateful applications · Data persistence · Filesystem layers · Pods · Namespaces · kubectl · Labels · Annotations · Service discovery · ReplicaSets · DaemonSets · Jobs · ConfigMaps · Secrets

Containers are lightweight, standalone software packages that include everything needed to run a piece of software, from the code to runtime, libraries, and dependencies [73]. Containers isolate applications from their environment, ensuring they operate uniformly regardless of the infrastructure [74]. This makes them ideal for microservices and scalable, cloud-native applications, where portability and efficiency are key. A **container image** is a snapshot of a specific application and its required environment. It encapsulates all dependencies and configurations required to run that application in a predictable, isolated way. Stored in a binary format, container images are designed to be rapidly deployed, and they are

B. Kumar et al., *Modern Kubernetes: From Core Concepts to Intelligent Autoscaling for Cloud Applications*, Studies in Autonomic, Data-driven and Industrial Computing, https://doi.org/10.1007/978-3-032-12972-7_6

typically managed and distributed through a **container registry**—a repository where images can be stored, tagged, and shared. The **Docker image format** has become the de facto standard for container images, largely due to Docker's role in popularizing container technology [75]. As containerization continues to evolve, the **Open Container Initiative (OCI)** has introduced the **OCI Image Format** to standardize container formats further. While this new standard is still maturing, it aims to foster compatibility across different container runtime environments, encouraging broader adoption in various systems and cloud environments. This chapter provides an in-depth analysis of Kubernetes version evolution, covering key milestones from foundational versions (v1.14–v1.20) to advanced capabilities (v1.26–v1.30). It examines ecosystem alignment, future roadmap developments, and strategies for smooth upgrades. Additionally, it explores stateful applications, container filesystem layers, and Kubernetes' role in orchestration. Practical insights into kubectl usage, namespaces, pods, labels, annotations, and service discovery are included. The chapter also covers scaling with ReplicaSets, and DaemonSets, and managing persistent data, highlighting challenges and strategies for ensuring data integrity in Kubernetes environments.

6.1 Understanding Filesystem Layers in Containers

Containers utilize a unique approach to handling filesystem data by relying on layered filesystems. Each container image is composed of a sequence of filesystem layers, with each layer representing incremental changes to the image's content. When a new container is created, an overlay filesystem merges these layers to provide a unified filesystem view [76]. This layering approach offers several advantages. It improves efficiency, as layers are reusable and shared across different containers, reducing redundancy and storage requirements. Additionally, it supports version control, enabling changes to the container image to be tracked layer by layer, which facilitates version management and rollback. The primary filesystems that facilitate this overlay approach include **aufs**, **overlay**, and **overlay2**. While the technical details of these systems vary, they all aim to efficiently combine multiple layers of filesystems, allowing changes to be isolated while maintaining base layers intact. Further exploration of these filesystems can reveal their specific performance characteristics, suitability for different environments, and compatibility with various container runtimes.

6.1.1 Configuration and Isolation of Container Environments

Container images are typically paired with configuration files that specify the parameters under which the containerized application should run. Key configuration concerns include **networking**, **namespace isolation**, **resource constraints**, and **syscall restrictions** [77]. Containers are designed to operate within isolated networks, which can be customized to control communication between containers or with external systems. Namespace isolation creates separate execution environments within the operating system, ensuring that containers remain isolated from each other, which is critical for security and resource management. Resource constraints are implemented using control groups (cgroups), enabling containers to be limited in their access to CPU, memory, and other system resources, thereby preventing them from exceeding their designated share. Additionally, syscall restrictions, enforced through security policies such as seccomp, limit access to certain system calls, providing an extra layer of security by reducing the attack surface. These configurations define how each container operates within a larger system, allowing administrators to create secure, efficient environments that maximize resource utilization without compromising isolation.

6.1.2 System Containers vs. Application Containers

Containers generally fall into two primary categories. **System containers** are designed to emulate an entire system, providing a near-complete OS environment within the container [78]. They serve as foundational containers upon which other application containers can be built and are particularly useful for applications requiring complex system dependencies or for creating test environments that simulate an entire operating system. On the other hand, **application containers** are intended to run a single application or service, typically layered on top of a base system container [79]. They are lightweight and optimized for quick deployment and scalability, making them particularly suited for microservices architectures, where each container hosts an isolated service. This design enables rapid scaling and modular development, which are key advantages in modern software deployment.

6.1.3 Kubernetes and Its Role in Orchestrating Containers

Kubernetes, an open-source container orchestration platform, provides essential services for managing, scaling, and deploying containerized applications across a cluster of machines [80]. With Kubernetes, containerized applications are no longer restricted to individual machines but are treated as units of computation that can be distributed across multiple nodes [81]. Among its key services, Kubernetes includes a load balancer, which automatically distributes incoming network traffic across multiple containers, ensuring an even distribution of application load. This improves availability and prevents any single container from becoming a bottleneck [82]. Additionally, Kubernetes features an internal DNS service for service discovery, enabling containers to communicate using predefined DNS names. This simplifies network configurations and makes services within the cluster easily accessible. For administrators, Kubernetes offers a web-based dashboard that provides real-time insights into cluster status, container health, resource usage, and other metrics, offering an intuitive interface for monitoring and managing the cluster.

6.1.4 Real-Life Example: Deploying a Web Application on Kubernetes

Consider a scenario where an organization wants to deploy a web application with a frontend, backend, and database service. Using Docker, each component (e.g., the frontend, backend, and database) is packaged into its container. These containers are stored as images in a container registry, accessible to Kubernetes for deployment. With Kubernetes:

1. Each component is deployed as a **pod** (the smallest deployable unit in Kubernetes), consisting of one or more containers.
2. Kubernetes ensures **service discovery** and networking by assigning each pod a unique IP address and managing inter-pod communication.
3. **Load balancing** automatically distributes user requests across multiple instances of the frontend, backend, or database, scaling the application based on demand.
4. If a container fails or a new version needs to be deployed, Kubernetes automates **rolling updates** and **self-healing** by restarting failed containers and gradually rolling out new versions without downtime.

In this way, Kubernetes simplifies the deployment, management, and scaling of complex, containerized applications, ensuring high availability and resource efficiency across a distributed cluster.

Overall, Containers and clusters, especially with Docker and Kubernetes, have transformed application deployment and management, enabling flexible, scalable, and highly isolated environments. Through effective use of container images, overlay filesystems, and Kubernetes orchestration, applications can be deployed with minimal friction and maximum resilience, forming the backbone of modern, cloud-native infrastructure.

6.2 Managing Kubernetes with kubectl

kubectl is the command-line interface for interacting with a Kubernetes cluster [82]. It allows administrators and developers to manage resources, deploy applications, monitor cluster health, and automate various aspects of containerized environments [83]. Mastering kubectl is essential for effectively working with Kubernetes, as it provides the tools needed to create, inspect, update, and delete resources within a cluster.

6.2.1 Using Namespaces for Resource Organization

Kubernetes namespaces act as logical partitions within a cluster, effectively managing and isolating resources. They are particularly beneficial for segregating different environments—such as production, staging, and development—within the same cluster. By organizing resources into namespaces, administrators can create isolated environments without requiring separate clusters, thereby streamlining both management and resource utilization. For instance, a development namespace might include resources like development databases, experimental application versions, and temporary services, catering to the needs of developers and testers. In contrast, a production namespace would host stable and live application versions, with stricter access controls and a focus on minimizing downtime to ensure high reliability. To interact with resources within a specific namespace, the –namespace flag is used with kubectl. For instance:

```
kubectl --namespace="prod" get pods
```

This command lists all pods within the prod namespace. Setting the namespace in this way allows for targeted management of resources, reducing the risk of unintended changes to resources in other environments.

6.2.2 Configuring Contexts for Persistent Environment Management

Contexts in Kubernetes allow users to save configurations and apply settings more permanently, making it easy to switch between clusters and namespaces without specifying them in each command [84]. Contexts are defined within the kubeconfig file and combine cluster, user, and namespace information. They're particularly useful for users who manage multiple clusters or frequently switch between environments. To create a context, you use a command like:

```
kubectl config set-context dev-context --cluster=dev-cluster --
    namespace=dev --user=dev-user
```

Switching to this context then becomes simple:

```
    kubectl config use-context dev-context
```

Once a context is set, commands issued with kubectl automatically apply to the configured cluster, user, and namespace. This ensures that changes persist across sessions, reducing errors from frequently switching between environments.

6.2.3 Working with Kubernetes Objects

In Kubernetes, resources are organized into various types, known as objects. Objects are fundamental entities within the Kubernetes system, representing the state of the cluster. Examples include Pods, Services, Deployments, ConfigMaps, and PersistentVolumes. Each of these object types represents a particular resource type with specific characteristics and functions within the Kubernetes ecosystem. To list all instances of a particular resource type, you can use the following command pattern:

```
kubectl get [resource]
```

For example:

- kubectl get services will list all services in the current namespace.
- kubectl get pods will display a list of all pods, showing their current status, age, and other relevant information.

To view a specific object within a resource type, provide the object's name:

```
kubectl get pod [pod-name]
```

6.2.4 Output Formats and Data Extraction

By default, `kubectl get` provides a human-readable output, displaying essential information about each resource. However, Kubernetes resources are defined in YAML or JSON, and these formats can be used for more precise data extraction or programmatic processing. To retrieve an object's configuration in JSON or YAML, use the `-o` (output) flag:

```
kubectl get pod [pod-name] -o yaml
kubectl get pod [pod-name] -o json
```

This command will output the complete configuration of the specified pod, including all metadata and settings, which can be helpful for debugging or auditing purposes. For Unix-like command-line utilities that work well with piped output, you can suppress column headers using the

```
--no-headers option:
kubectl get pods --no-headers
```

This is particularly useful for commands that extract specific information, making it easier to parse output with other command-line tools.

6.2.5 Advanced Output Filtering with JSONPath

For scenarios where only certain fields from a resource's configuration are needed, `kubectl` provides filtering options through JSONPath, a query language for JSON that allows you to select specific data fields. This is especially useful for scripting or automation tasks where concise output is required. To use JSONPath with `kubectl`, specify a template with the `-template` option. For example, to extract the name and status of all pods:

```
kubectl get pods --template="{.items[*].metadata.name} {.
    items[*].status.phase}"
```

This command provides a structured output containing only the requested fields, which can then be piped into other tools for further processing.

6.2.6 Real-Life Example: Managing a Staging Environment

Consider a real-world scenario in which a team is managing a Kubernetes cluster for a web application with both staging and production environments. Using `kubectl` and namespaces, the team can segregate resources based on their environment, making management more intuitive and minimizing the risk of accidental changes.

1. Set up the namespaces:

```
kubectl create namespace staging
kubectl create namespace production
```

2. Deploying to Staging: For testing purposes, developers deploy the latest application version in the staging namespace:

```
kubectl --namespace=staging apply -f deployment.
    yaml
```

3. Switching Context: To prevent accidental changes to production, developers set a context for the staging environment and use it for all staging-related commands:

```
kubectl config set-context staging-context --
    cluster=my-cluster --namespace=staging --user
    =developer kubectl config use-context staging
    -context
```

4. Viewing Resource Status: With the staging context set, developers list the current state of resources:

```
kubectl get all
```

5. Extracting Specific Information: Developers extract only the pod names and statuses for the staging deployment, facilitating a quick health check:

```
kubectl get pods --namespace=staging --template="
    {.items[*].metadata.name {.items[*].status.
    phase}"
```

By leveraging namespaces, contexts, and JSONPath queries, the team manages complex multi-environment setups efficiently, minimizing the potential for errors and simplifying the deployment process.

Overall, the `kubectl` command-line tool is an indispensable part of Kubernetes management, providing granular control over clusters, namespaces, and resources. Administrators can effectively manage and automate their Kubernetes environments by organizing resources into namespaces, using contexts for persistent settings, and leveraging JSONPath queries for data extraction. These practices are essential for building scalable, reliable, and secure containerized applications.

6.3 Understanding and Managing Pods in Kubernetes

In Kubernetes, pods are the fundamental deployment unit, representing the smallest building block that can be created, managed, and orchestrated. A pod is a wrapper for one or more containers that share resources and are deployed together as a single unit [17]. Containers within a pod share a common IP address, port space, hostname, and interprocess communication capabilities. Although containers in a pod can run independently, they are treated as a single entity, making pod management more efficient for application deployment.

6.3.1 Shared Resource Within a Pod

Containers within the same pod share several system resources, enabling them to function cohesively as a single application component. For instance, they share an **IP address**, allowing seamless communication between containers without requiring additional networking configurations. Additionally, a shared **port space** simplifies inter-container communication within the pod by enabling containers to use predefined ports for interaction. They also share a **hostname**, managed through the UTS (UNIX Timesharing System) namespace, which streamlines naming conventions within the pod. Furthermore, **interprocess communication (IPC)** mechanisms, such as System V IPC and POSIX message queues, allow containers to share data and synchronize processes efficiently. These shared resources and mechanisms enhance the integration and functionality of containers within a pod, making them well-suited for building unified application components. Pods are the smallest deployable units in Kubernetes and are essential for managing workloads. Each pod is ephemeral; if it crashes, Kubernetes restarts it. Pods encapsulate a container or group of containers and are isolated from other pods as if they were on separate servers. This isolation is crucial for achieving a microservices architecture where each pod represents a distinct application component. Pod manifests are stored persistently in etcd, the key-value store that Kubernetes relies on for cluster data storage. When a pod is created, its specification is saved to etcd, ensuring consistency and allowing Kubernetes to restore the pod if necessary.

6.3.2 Managing Pod Lifecycle and Interactions in Kubernetes

Kubernetes provides robust mechanisms for managing pod lifecycles and facilitating interactions with pods, ensuring applications run reliably and efficiently. The pod lifecycle includes starting, monitoring, and stopping pods, with a focus on graceful termination. When a pod is terminated, Kubernetes initiates a default grace period of 30 seconds, during which it sends a SIGTERM signal to allow the application to complete its operations and save its state. If the pod has not shut down by the end of this period, Kubernetes forcibly terminates it using a SIGKILL signal, ensuring cluster resources are reclaimed. In addition to lifecycle management, Kubernetes offers tools to interact with pods, even if they are not publicly accessible. Port forwarding allows users to establish a temporary direct connection by forwarding traffic from a local port to a pod's port, enabling external access for testing or debugging purposes. This is achieved using the

```
kubectl port-forward [pod-name] 8080:8080
```

Another useful feature is file copying, where the Kubectl cp command facilitates transferring files between a local machine and a container within a pod. This capability is particularly valuable for troubleshooting or making on-the-fly configuration changes. Together, these features empower administrators and developers to efficiently manage and interact with pods in a Kubernetes cluster.

- Health Checks: Ensuring Pod Reliability: Kubernetes includes three main types of probes to monitor pod health and readiness:
  - Process Probe: This always-on probe monitors the main process within a pod. If this process crashes, Kubernetes restarts the pod. The process probe has an exponential backoff for restarting, ensuring the application does not repeatedly fail without intervention.
  - Liveness Probe: Configured in the pod manifest using YAML, a liveness probe checks if the pod is active and functioning. If the pod fails this check, Kubernetes terminates and recreates it. This probe can be defined using several methods:

    HTTP/S Requests: Sends an HTTP request to a specific endpoint.

TCP Socket Requests: Opens a TCP connection to a defined port.
Exec Scripts: Runs a script inside the container to determine health based on the exit code.

- Readiness Probe: This probe checks if a pod is ready to serve traffic. If a pod fails the readiness probe, it is temporarily removed from the load balancer rotation until it passes. Readiness probes are useful for ensuring that only healthy pods receive traffic, enabling a smoother user experience

6.3.3 Resource Requests and Limits

Each container within a pod can have defined resource requests and limits for CPU, memory, and other system resources. These settings are crucial for ensuring fair resource distribution and optimizing utilization across the cluster. Requests specify the minimum guaranteed resource allocation for a container, ensuring it has sufficient resources to operate effectively. Limits, on the other hand, define the maximum resources a container can consume if additional capacity is available. For example, if a pod contains two containers, each requesting 500m (0.5) CPU cores, Kubernetes initially allocates this guaranteed amount. While CPU usage can burst beyond the requested amount if the node has unused capacity, memory usage strictly adheres to the specified limits, preventing over-consumption and potential disruptions.

6.3.4 Disk Storage and Volume Mounting

Kubernetes provides various volume types to support data storage within pods, catering to different use cases: **emptyDir** creates a temporary directory that lasts as long as the pod is running, making it ideal for caching or temporary data storage. **Persistent Volumes** enable connections to external storage solutions, such as NFS or cloud storage services like Amazon S3, ensuring data persists beyond the pod's lifecycle. **HostPath volumes** grant access to storage on the underlying node; however, this approach is discouraged in production environments due to the risk of data inconsistencies if the pod is rescheduled to a different node. Volumes are mounted at the container level within a pod, allowing each container to access one or more volumes. These volumes can be mounted at distinct paths, facilitating flexible and diverse storage configurations within the same pod.

6.3.5 *Real-Life Example: Web Application with Multiple Containers in a Pod*

In Kubernetes, pods provide a cohesive environment for deploying containers that need to work closely together, such as a web application with a frontend and backend. By deploying both containers in a single pod, the frontend and backend share the same IP address and network resources, enabling efficient communication as if they were running on the same server. Each container is assigned specific responsibilities, with resource configurations tailored to their requirements—for instance, the frontend might request fewer CPU and memory resources compared to the backend, which typically has higher processing demands. To ensure application reliability, liveness and readiness probes are configured: the frontend might use an HTTP liveness probe to check the main page's availability, while the backend could employ a readiness probe to verify database connections before processing requests. Additionally, a shared storage volume, such as an emptyDir volume, can be mounted to facilitate temporary data exchange between the frontend and backend, ensuring seamless user session management. By understanding and leveraging these features, pods act as the foundational unit for deploying and managing containerized applications, offering shared resources, well-defined lifecycles, and high configurability to achieve reliability, availability, and performance in a Kubernetes cluster.

6.4 Labels and Annotations in Kubernetes

In Kubernetes, labels and annotations are metadata components that add flexibility and insight into resource management. They allow users to categorize, organize, and manage Kubernetes objects effectively, accommodating the ever-evolving understanding of system architecture. While labels are intended primarily for object identification and grouping, annotations serve as data storage for external tools and configurations.

6.4.1 *Labels: Identifying and Grouping Kubernetes Objects*

Labels are arbitrary key-value pairs that can be applied to Kubernetes objects like pods, services, and deployments. They are the foundational components for grouping and managing Kubernetes resources. Using labels, users can define meaningful categories for organizing system components, helping streamline resource selection and filtering.

- Key-Value Pairs: Each label consists of a key and a value, which can represent any identifier for a Kubernetes resource. This flexibility allows users to group objects by attributes such as environment (prod, staging), version (v1, v2), or app (frontend, backend).

- Label Prefixes: Labels may optionally include a prefix in DNS format, e.g., app.example.com/env. This prefix helps avoid naming conflicts, especially in shared or multi-tenant environments.
- Assignment of Labels: Labels can be applied when defining a Kubernetes object in YAML, or they can be added later using kubectl commands:

```
kubectl label deployments my-deployment "env=
    production"
```

- Removing Labels: To delete a label, append a - after the key-value pair:

```
kubectl label deployments my-deployment "env-"
```

6.4.2 Using Label Selectors for Object Management

Kubernetes employs label selectors as a versatile mechanism for filtering and retrieving information about labeled objects. These selectors allow users to apply complex criteria to manage resources efficiently. Command-Line Selectors can be used with the –selector flag in kubectl get commands to filter objects based on their labels. For example, the command

```
kubectl get pods --selector="env=production,app=
    frontend"
```

retrieves pods labeled with both env=production and app=frontend. In addition to command-line usage, Template Selectors are employed in YAML or JSON configurations. These selectors utilize matchExpressions to define boolean combinations within dictionaries, providing an intuitive way to specify conditions for resource selection [85]. This approach is commonly used in configurations for ReplicaSets and Deployments, enabling precise control over the resources managed by Kubernetes. This syntax is commonly seen in ReplicaSet and Deployment configurations:

```
selector:
    matchExpressions:
        - key: "env"
        operator: In
        values: ["production", "staging"]
```

By leveraging these label selectors, administrators and developers can streamline resource management and ensure that operations target the desired objects effectively.

6.4.3 Annotations: Storing Non-identifying Metadata

While labels help organize and filter resources, annotations serve as a storage mechanism for non-identifying metadata. Annotations provide a way to attach extra information to objects for external tools, without impacting how Kubernetes manages the object.

- Arbitrary Data Storage: Annotations are key-value pairs with no restriction on the length or complexity of the value, allowing storage of data blobs or configuration information.
- Metadata Purposes: Annotations are often used for storing data needed by tools that work with Kubernetes but aren't relevant to the Kubernetes API directly, such as monitoring configurations or additional descriptive information.
- Usage in YAML Manifests: Annotations are specified under the metadata section of an object definition:

```
metadata:
    name: my-pod
    annotations:
      description: "This pod is used for demo purposes"
      prometheus.io/scrape: "true"
```

6.4.4 Labels vs. Annotations: When to Use Each

Labels and annotations serve distinct purposes in Kubernetes resource management, despite their similarities. **Labels** are used for identification, selection, and grouping of resources. They play a crucial role in Kubernetes' scheduling and selection logic, making them indispensable for controlling resource behavior. In contrast, **annotations** are designed to store non-identifying metadata. Unlike labels, annotations do not influence Kubernetes' selection logic and are better suited for providing auxiliary information intended for external systems or tools. In practice, labels should be employed when filtering or organizing resources is required, while annotations are ideal for storing supplementary data that does not affect Kubernetes' object handling.

- **Real-World Example:** Using Labels and Annotations for Application Management Consider an application deployed across different environments (production and staging) with multiple versions. Here's how labels and annotations can be used effectively as using Labels for Identification:
  - The env label categorizes resources by environment.
  - The version label identifies the application version.

– This setup allows users to select all production pods for version 1:

```
kubectl get pods --selector="env=production,
    version=v1"
```

and Annotations for Metadata:

– Each pod has an annotation providing a description for easy identification in external monitoring.
– The prometheus.io/scrape annotation instructs monitoring tools like Prometheus to collect metrics from the pod.

Overall, labels and annotations offer flexible ways to manage Kubernetes resources and provide metadata that supports system observability and configuration [86]. Labels drive core Kubernetes functions, such as filtering and grouping, while annotations store detailed metadata for use by external systems [87]. By understanding how to use both effectively, users can better organize, observe, and extend Kubernetes-managed applications.

6.5 Service Discovery in Kubernetes

Service Discovery in Kubernetes enables applications within a cluster to locate and communicate with one another, even as they scale or shift across nodes [88]. In Kubernetes, each Service acts as an abstraction layer over a group of pods, creating a stable network endpoint that allows reliable connectivity regardless of pod IP changes. Kubernetes supports two primary methods for service discovery: ClusterIP and DNS-based discovery. With ClusterIP, Services within the same cluster can communicate via an internal IP address, while DNS-based discovery leverages Kubernetes' DNS service to automatically assign DNS names to Services, making it easier for applications to locate each other by name. This model allows seamless scaling and flexibility, helping maintain application availability and efficient communication across complex, dynamic infrastructures.

1. The Challenge of Naming and Service Discovery in Kubernetes: Service discovery—figuring out how services within a cluster can find and connect to each other—is one of the foundational challenges in Kubernetes. Traditional DNS, as used on the Internet, is not ideal for the dynamic and high-frequency changes within Kubernetes clusters. Issues like TTL (time-to-live) settings, round-robin balancing, and handling high numbers of A records make conventional DNS ill-suited for Kubernetes's needs. Kubernetes addresses these challenges by using a specialized DNS setup integrated with its Service object API.
2. The Service Object: A Foundation for In-Cluster Networking: Kubernetes introduces **Service objects** to streamline internal service discovery, simplifying how deployments within a cluster locate and interact with one another. Each Service

object provides a stable virtual IP address, known as a **Cluster IP**, which acts as a router rather than a physical interface. The **Cluster IP** is permanent within the cluster's DNS and can be safely cached indefinitely, as it remains unchanged even if the underlying pods are redeployed. When deploying an application, using the kubectl expose command creates a Service object and automatically registers it within the cluster's DNS. Other pods can leverage Kubernetes' built-in DNS service to resolve the Cluster IP and port associated with any deployed Service. This architecture enables seamless, low-latency communication within the Kubernetes network, ensuring that each Service is accessible exclusively within the cluster.

3. Virtual IPs and Internal Routing: The Cluster IP address acts as a stable and reliable access point within the Kubernetes cluster, efficiently directing traffic to the appropriate pods. This design optimizes in-cluster routing by keeping traffic local, thereby minimizing network transit latency. However, the Cluster IP restricts service accessibility within the cluster, limiting external access. To enable external accessibility, Kubernetes provides additional configuration options such as **NodePort** and **LoadBalancer**. **NodePort** exposes a service on a specific port on each node, allowing external traffic to be routed from the node to the Cluster IP and then to the appropriate pods. This mechanism ensures that external users can access services within the cluster by connecting to the specified port on any node. On the other hand, **LoadBalancer** introduces a more sophisticated abstraction layer for traffic distribution. Its configuration varies across cloud providers, offering flexible ways to balance traffic across multiple nodes, thereby improving performance and availability for external access.
4. DNS Naming in Kubernetes: Kubernetes uses a hierarchical DNS naming convention for services within a cluster. An example DNS name might be using Arduino:

```
alpaca-prod.default.svc.cluster.local
```

The components of this name provide context about the service:

- Service name (alpaca-prod): Specifies the name of the Service.
- Namespace (default): Identifies the namespace where the service is running.
- Service indicator (svc): Labels the entity as a Kubernetes service.
- Cluster domain (cluster.local): Defines the base domain name for the cluster. This can be customized by administrators but is mostly cosmetic.

Using this DNS-based naming, applications can request a service by sending HTTP requests directly to its service name, ensuring smooth communication within the cluster without manual configuration.

6.5.1 Endpoint Objects: Direct Communication with Pods

For applications requiring direct interaction with individual pods, such as avoiding load balancing or maintaining sessions, Kubernetes provides Endpoint objects. These objects store the specific IP addresses of the pods behind a service, enabling clients to communicate directly with particular instances. By leveraging **direct pod IP access**, Kubernetes APIs allow users to retrieve and subscribe to updates on individual pod IPs, effectively bypassing DNS-based service discovery. This approach can enhance performance for Kubernetes-native applications by eliminating the overhead of DNS lookups and load balancing, ensuring more efficient and predictable communication.

6.5.2 Methods and Best Practices for Service Discovery

In Kubernetes, service discovery can be managed either manually or automatically, with each approach having its own advantages and challenges. Developers can manually discover services by querying the Kubernetes API for specific pod IP addresses. However, this method can be difficult to synchronize across rapidly changing resources, which is why Kubernetes Service objects are the preferred way to handle pod-to-pod communication. The **kube-proxy** service plays a critical role in this process, managing networking and routing rules within the cluster. It dynamically configures the operating system's iptables to route traffic to and from services, maintaining virtual IPs, NodePorts, and other networking elements. Additionally, while it is possible to use environment variables to manually stitch services together, this approach is generally discouraged due to the need for strict control over the service startup order, which can lead to errors. Relying on environment variables for service discovery can also undermine reliability and scalability. In contrast, using Kubernetes **Service objects** and **DNS-based service discovery** offers a more resilient, scalable solution, as services are automatically routed to the correct resources, reducing the chances of misconfiguration and enhancing infrastructure flexibility.

Overall, Service discovery in Kubernetes enables seamless, resilient communication between services, optimized for the cluster's dynamic environment. By leveraging Service objects, Cluster IPs, and kube-proxy's routing management, Kubernetes simplifies internal service discovery while providing flexibility for external access when needed. Using Kubernetes' built-in DNS and API-based methods allows developers to design highly interconnected applications without manual networking configurations

6.6 ReplicaSets in Kubernetes

A ReplicaSet is a fundamental Kubernetes resource designed to manage the desired number of pod replicas in a cluster, ensuring both resilience and scalability. Unlike Service objects, which handle service discovery and pod accessibility, ReplicaSets focuses on maintaining the specified pod count and enabling self-healing. By managing a set of identical pods based on a single container image, ReplicaSets ensures that applications remain highly available. If a pod crashes or becomes unresponsive, the ReplicaSet automatically recreates it, maintaining the desired state and resilience. Additionally, ReplicaSets support seamless scale-out by adjusting the number of replicas as needed. While ReplicaSets can create and manage pods directly, best practices in Kubernetes recommend using Deployments to oversee ReplicaSets, as this approach provides greater control and simplifies scaling and updates, enhancing operational efficiency.

6.6.1 Scaling with ReplicaSets

Kubernetes provides two methods for scaling ReplicaSets:

- Declarative Scaling: The recommended method is to update the source file that defines the ReplicaSet (or Deployment) and use kubectl to apply the changes. This keeps the cluster state in sync with the files defining it, making the system easier to manage and troubleshoot.
- Imperative Scaling: Alternatively, the kubectl scale can be used to directly alter the number of replicas. However, this approach can cause desynchronization between the cluster's live state and its configuration files, making debugging more challenging.
- Automatic Horizontal Scaling: For horizontal scaling based on metrics like CPU or memory usage, Kubernetes offers the kubectl autoscale command. This command configures automatic scaling for the ReplicaSet, adjusting the number of replicas based on real-time usage data.
- No Vertical Scaling: Currently, Kubernetes lacks a built-in solution for vertical scaling (i.e., adding resources like CPU and memory to existing pods). ReplicaSets are therefore focused solely on horizontal scaling to handle increased demand by adding pod instances rather than increasing the resources of existing pods.

6.6.2 Working with Deployments, Services, and ReplicaSets

In a typical Kubernetes deployment workflow, the process begins with creating a Deployment configuration file (e.g., deployment.yml), which specifies the ReplicaSets responsible for managing the application's pods. Once the configuration is prepared, it is applied using the kubectl apply -f deployment.yml command, launching the Deployment. This command automatically manages the associated ReplicaSet(s) and the pods they control. Afterward, the deployment is exposed to the cluster's network by running the kubectl expose deployment command, which creates a Service object. This Service object facilitates service discovery by registering the deployment with Kubernetes DNS, making it accessible to other pods. By following this workflow, Kubernetes ensures that all necessary components—such as pods, ReplicaSets, and Services—are properly configured, enabling a reliable and scalable deployment.

6.6.3 ReplicaSets and Deployment Boundaries

When designing deployments in Kubernetes, it's important to consider boundaries based on team responsibilities and logical separations. Deployments should generally manage ReplicaSets that are relevant only to their own functionality, and pods managed by other teams or unrelated components should not be included within the same Deployment. This boundary management simplifies both operations and debugging by ensuring clear responsibility for each Deployment's lifecycle.

Overall, ReplicaSets are a crucial part of Kubernetes' scaling and resilience toolkit, ensuring that applications have the correct number of running pods and can recover from failures. By managing ReplicaSets through Deployments and exposing services with Service objects, Kubernetes provides a robust framework for deploying scalable and resilient applications. While automatic horizontal scaling is supported, Kubernetes users should be aware of limitations with vertical scaling and be mindful of using declarative methods to keep configuration files in sync with the live cluster state.

6.7 DaemonSets in Kubernetes

A DaemonSet in Kubernetes is similar to a ReplicaSet but is designed for use cases where certain pods need to run on every node in the cluster. DaemonSets are ideal for system-level applications such as log collectors, monitoring agents, and networking services that require a presence on each node to function effectively across the cluster.

6.7.1 Key Characteristics of DaemonSets

- Uniform Deployment Across Nodes: DaemonSets ensure that a single instance of a pod is scheduled on every node, making it perfect for cluster-wide services.

- Selective Node Deployment: You can specify certain nodes for the DaemonSet to deploy pods on, using label selectors to target specific node groups. This flexibility enables you to deploy DaemonSets only on nodes that meet specific criteria.
- Automatic Scaling to New Nodes: When a new node is added to the cluster, the DaemonSet automatically deploys the required pod instance on it, ensuring consistent availability of the service across all nodes.

6.7.2 Rolling Updates and Use Cases for DaemonSets

DaemonSets in Kubernetes are designed to ensure that a specific pod runs on every node in the cluster, providing essential infrastructure services like log collection, monitoring, and networking. One of the key features of DaemonSets is their ability to support rolling updates, though this capability is not enabled by default for backward compatibility. To implement rolling updates, you must explicitly configure the update strategy within the DaemonSet's configuration file. This allows you to update the pods incrementally, replacing one pod at a time, ensuring that the service remains available throughout the update process, avoiding disruptions to service across the cluster. DaemonSets are particularly well-suited for use cases where a consistent presence of pods is required on all nodes in the cluster. For instance, they can be used for log collection, such as deploying log collectors like Fluentd across all nodes to gather logs from applications running on the cluster. Similarly, DaemonSets are ideal for deploying monitoring agents, such as Prometheus node exporters, which collect node-specific metrics, or for deploying networking tools that manage network traffic uniformly across the nodes in the cluster. These applications benefit from DaemonSets' ability to ensure that essential infrastructure components are available on every node in the cluster.

Overall, DaemonSets are essential for deploying services that need uniform presence across all nodes in a Kubernetes cluster. By ensuring that a single instance of a pod is automatically scheduled on every node, DaemonSets simplifies the management of cluster-wide services. Although rolling updates are supported, they require explicit configuration, giving you control over how updates are managed without disrupting the service's availability.

6.8 Jobs in Kubernetes (Job Objects)

In Kubernetes, a Job object is designed to manage one-time tasks that need to run until completion. Unlike regular pods, which might continuously restart upon exiting, Jobs allows you to create pods that run to completion and terminate after completing their tasks. This makes them ideal for tasks like batch processing, script executions, and other non-continuous workloads.

- Key Features of Jobs:
  - One-Time Execution: A Job creates pods that execute a defined task and then terminate. If you attempted to run an existing container as a standard pod, it would restart indefinitely, causing issues. Jobs ensure that once a pod completes its task, it doesn't automatically restart.
  - Customizable Restart Policy: You can control the restart behavior of a Job by specifying –restart=Never or –restart=OnFailure, allowing you to decide if failed tasks should be retried.
  - Command Line Arguments: Jobs support passing arguments directly to the container's ENTRYPOINT using the – flag, making it easier to customize the job's behavior without modifying the container image.
- Running a Job: To launch a Job, you can use the kubectl run command with specific flags to configure its behavior:
  - –restart=Never: Ensures the Job runs to completion without retrying after termination.
  - -i (Interactive): Streams the logs from the first pod in the job to the terminal's standard output (STDOUT).
  - – [args]: Passes custom command-line arguments to the container's ENTRYPOINT, which allows you to control the container's behavior for each Job run.
- Job Execution Patterns: Kubernetes offers different job types to handle varying workload requirements. **One-Shot Jobs** is designed to execute a single task or script, akin to a one-time script execution, but managed at scale within Kubernetes. Once the task is completed, the Job terminates automatically. For more complex tasks that can be divided into smaller, independent units, **Parallelized Jobs** come into play. These jobs allow you to configure two key parameters—parallelism and completions. Parallelism specifies the number of pods that can run concurrently, while completions define the total number of successful completions needed. Kubernetes automatically manages the creation of new pods to fulfil the completion requirement, improving efficiency. For example, setting parallelism to 3 and completions to 9 means that up to 3 pods can run concurrently, with new pods being created until 9 successful completions are reached, accelerating the processing of large workloads. Lastly, **Work Queue Jobs** are ideal for continuously processing tasks from a work queue without needing a fixed number of completions. In this setup, a parallelism value is specified, but the completion count is omitted, ensuring that pods continue processing tasks until all work is completed or a pod exits successfully. This flexible approach is particularly useful for dynamic, ongoing workloads.

Overall, Jobs in Kubernetes provides a robust way to manage one-time and parallel workloads. By defining specific parameters, you can control the concurrency, retry behavior, and completion criteria, making Jobs suitable for a wide range of tasks, from simple scripts to large-scale parallel computations. With options for passing custom arguments and managing output, Kubernetes Jobs allow efficient and flexible management of task-based workloads within your cluster.

6.9 ConfigMaps and Secrets in Kubernetes

In Kubernetes, managing configuration data and sensitive information securely and efficiently is crucial for building scalable and maintainable applications. Two essential resources that serve these purposes are **ConfigMaps** and **Secrets**. ConfigMaps allows developers to manage non-sensitive configuration data, such as environment settings and application parameters, in a flexible and reusable manner. These configurations can be shared across multiple containers, making them ideal for applications that require consistent environment settings without hardcoding values. ConfigMaps provides a mechanism for injecting non-sensitive configuration data into Kubernetes pods. They offer three primary ways to pass data into containers: first, by mounting configuration data as files, which are directly accessible from the container's filesystem; second, by setting environment variables, enabling applications to access configuration at runtime; and third, by injecting values directly into command-line arguments, allowing dynamic configuration changes without modifying the container image. However, it's important to note that while ConfigMaps offer flexibility, they do not automatically update containers when their values change. Applications depending on ConfigMaps may need to be manually restarted or reloaded to pick up updates. In contrast, Kubernetes **Secrets** are specifically designed to manage sensitive information, such as passwords, tokens, and private keys, securely. Unlike ConfigMaps, Secrets are encrypted when stored and are only accessible in a controlled way within the container. Secrets are mounted as files within containers, and placed in an in-memory (tmpfs) RAM partition, ensuring that the sensitive data is not written to disk and remains secure. This method of exposure ensures that the sensitive data stays confidential and is only available to the application as needed, providing a high level of security for sensitive information.

Together, ConfigMaps and Secrets in Kubernetes enable developers to manage both configuration data and sensitive information in a secure and flexible manner. While ConfigMaps offer a way to handle non-sensitive data across containers, Secrets ensures that critical credentials are managed securely. By using both, Kubernetes applications can achieve a high level of security, flexibility, and maintainability in configuration management.

- Special Case: Image Pull Secrets An image pull secret is a specific type of Secret used for authenticating Kubernetes with private Docker registries. It allows Kubernetes to pull images securely from a private repository. To create an image pull secret, use:

```
    kubectl create secret docker-registry <secret-name>
  --docker-username=<username>
  --docker-password=<password>
  --docker-email=<email>
```

After creating the secret, you can reference it in the imagePullSecrets field of your pod or deployment definition.

Overall, ConfigMaps and Secrets help manage configuration and sensitive data, respectively, enabling secure, configurable, and portable applications in Kubernetes. While ConfigMaps allow for flexible configuration injections, Secrets provides controlled access to confidential information. Both ConfigMaps and Secrets improve modularity and security, making them essential tools for managing configurations in a Kubernetes environment [89].

6.10 Deployments in Kubernetes

In Kubernetes, a Deployment is an object designed to manage the lifecycle of applications by deploying and updating ReplicaSets and Pods. Deployments provide robust capabilities for rolling updates and scaling, making them essential for maintaining the high availability and reliability of applications [90]. With Deployments, Kubernetes ensures that applications are always running the desired number of Pods and that any updates or changes to the application are handled efficiently. Kubernetes supports two main deployment strategies for managing how applications are updated and rolled out. So first, **RollingUpdate** (default) strategy ensures minimal downtime by gradually replacing old Pods with new ones. The update process is carefully managed to maintain a balance between old and new versions, allowing for a seamless transition that minimizes service disruption. The RollingUpdate strategy is controlled using several key parameters:

- **maxUnavailable**: Specifies the maximum number of Pods that can be unavailable during the update process, ensuring that the application remains functional even as it is updated.
- **maxSurge**: Defines the maximum number of extra Pods that can be created above the desired count during an update, allowing Kubernetes to temporarily scale up the application during the update.
- **minReadySeconds**: Ensures that each updated Pod is in a "Ready" state for a minimum duration before proceeding with the next update, preventing disruptions caused by premature pod replacements.
- **progressDeadlineSeconds**: Sets the maximum duration that Kubernetes will wait for an update to make progress before considering it a failure, helping to detect stalled updates.

These parameters provide fine-grained control over how the update process occurs, allowing administrators to balance the need for minimal downtime with the desire to roll out new application versions quickly.

The **Recreate** strategy, on the other hand, terminates all existing Pods before creating new ones. While this approach can lead to service downtime, as no Pods are running during the transition, it may be appropriate for certain use cases where downtime is acceptable or for experimental environments. However, this strategy is generally discouraged in production environments where high availability is required. **Best Practices for Deployments:** To ensure a reliable deployment process, it is recommended to use a declarative configuration approach. This involves editing a configuration file and applying it rather than making imperative changes directly. By doing so, you take advantage of Kubernetes's reconciliation loop, which continuously checks for and applies the desired configurations, ensuring consistency and reliability. Additionally, it is essential to ensure backward compatibility during rolling updates. Your application should be able to handle both old and new data formats simultaneously, preventing potential issues if services rely on both versions during the transition. By following these best practices and utilizing Kubernetes's deployment strategies, teams can effectively manage application updates and maintain high availability throughout the service lifecycle.

6.11 Managing Persistent Data in Kubernetes

While Kubernetes excels at managing stateless applications, handling persistent data can be a complex challenge [91]. Stateless applications can be easily managed with containers, as they do not rely on stored data between requests [92]. However, most real-world applications need to persist state across sessions, which introduces a set of challenges when running in a containerized environment.

6.11.1 Key Challenges with Persistent Data

1. No Clean Start: Unlike stateless applications, persistent data often already exists somewhere (e.g., on a disk or in a database). Managing this data may require migration or adaptation to Kubernetes, which is not always straightforward.
2. Non-Swappable Storage: Storage devices containing data cannot be simply swapped with identical empty devices. Kubernetes does not natively support hot-swapping of persistent storage.
3. Kubernetes Focus on Stateless: The Kubernetes API is optimized for stateless applications. Working with persistent data often feels "clunky" as Kubernetes is not inherently designed for managing complex, stateful services.

6.11.2 Strategies for Managing Persistent Data

1. **Using External Storage with Kubernetes Services:** A recommended approach is to treat external storage (such as an existing database or other services) as a Kubernetes Service object. This abstracts away the underlying infrastructure and allows for easy integration with Kubernetes-managed applications. The external storage can be exposed using the externalName field, which points to the external service's CNAME (or IP address if no CNAME exists). This makes it easy to

replace external services with internal ones when necessary, especially useful for testing.

2. **PersistentVolume and PersistentVolumeClaim:** A more Kubernetes-native approach involves using PersistentVolumes (PVs) and PersistentVolumeClaims (PVCs). A PVC allows a pod to request storage resources, which can then be dynamically provisioned by the cloud provider. While simpler than managing external services, this method does not inherently provide high availability (HA), as it relies on a single persistent volume.
3. **StatefulSets for Advanced Persistent Storage:** StatefulSets are an advanced feature in Kubernetes for managing stateful applications [93]. They are similar to ReplicaSets but come with additional features to support persistence [94]. StatefulSets provides the following:
   - Unique, Stable Hostnames: Each replica in a StatefulSet is assigned a unique, persistent hostname (e.g., db-0, db-1, etc.), which ensures that the stateful application retains its identity across restarts.
   - Ordered Deployment and Scaling: Pods in a StatefulSet are created in a specific order (low-to-high), and scaled down in reverse order (high-to-low). This ensures that the applications deployed in these pods are aware of each other and can maintain consistency.
   - Headless Services: A StatefulSet typically uses a "headless" Service (i.e., with clusterIP: None) to expose individual pods directly, rather than using a load-balanced IP address. This setup allows applications to use DNS to resolve the specific pod IPs, making it ideal for stateful applications that need direct access to specific instances.
4. PersistentVolumeClaim Templates for StatefulSets: To handle storage for the pods managed by StatefulSets, Kubernetes allows the use of PersistentVolume-Claim templates. This ensures that each pod in the StatefulSet can claim its own persistent volume dynamically. However, it's important to ensure that there is enough storage capacity available in the cluster to accommodate the StatefulSet's storage requirements. If not, the StatefulSet will fail to deploy due to insufficient resources.

Managing persistent data in Kubernetes requires understanding the unique challenges and leveraging the appropriate strategies for different use cases. While external services can be integrated via Kubernetes Service objects, more complex stateful applications can benefit from using StatefulSets and PersistentVolumes. By carefully selecting the right approach based on your application's needs, you can effectively manage persistent data in a containerized environment, ensuring both reliability and scalability.

6.12 Summary

Kubernetes has evolved into a powerful tool for managing containerized applications at scale, primarily excelling in stateless workloads. However, as stateful applications become more common in modern environments, the need to address data persistence, consistency, and availability within Kubernetes has grown. This book has explored the various tools and resources Kubernetes offers for managing stateful applications, such as **ReplicaSets**, **DaemonSets**, **Jobs**, **Deployments**, **ConfigMaps**, **Secrets**, and **StatefulSets**. Each of these resources plays a critical role in maintaining application availability, ensuring data persistence, and managing complex data scenarios. The integration of **PersistentVolumes** and **PersistentVolumeClaims** has allowed Kubernetes to handle persistent storage effectively, with dynamic provisioning capabilities provided by cloud providers. **StatefulSets** offers specialized support for applications that require stable network identities and persistent storage, ensuring that complex stateful applications can be reliably managed. Through this detailed exploration, Kubernetes users can now confidently approach the challenges of managing stateful workloads in a Kubernetes environment. With the right understanding of **Services**, **Persistent Storage**, and Kubernetes-native features like **ConfigMaps** and **Secrets**, users can ensure that their applications maintain high availability, scalability, and data consistency. The advanced topics, including data migration and high availability, provide additional insights into managing complex storage requirements and ensuring that Kubernetes remains a flexible, robust platform for both stateless and stateful applications. Ultimately, the tools and strategies discussed in this book offer Kubernetes users the knowledge necessary to handle even the most challenging stateful application scenarios, making Kubernetes a versatile and reliable platform for modern cloud-native application management.

6.13 Review Questions

- Container Fundamentals
  - How do filesystem layers in containers differ from traditional VM storage?
  - Compare system containers vs application containers in Kubernetes environments.
  - Explain how Kubernetes orchestrates containers differently than standalone Docker.
- kubectl and Cluster Management
  - What kubectl command creates a new namespace for staging environments?
  - How would you switch contexts between production and development clusters?
  - Demonstrate JSONPath filtering to extract only Pod IP addresses from kubectl output.

  - What are three common Kubernetes object types and their management commands?
- Pod Management
  - Why would you deploy multiple containers in a single Pod?
  - Explain how resource requests/limits prevent "noisy neighbor" problems in Pods.
  - What volume type would you choose for a Pod needing temporary scratch space?
  - How does Kubernetes handle container restarts within a Pod?
- Labels and Annotations
  - Write a label selector query to match all "production" environment Pods with "web" tier.
  - When would you use annotations instead of labels for metadata?
  - How do labels enable service discovery in Kubernetes?
- Service Discovery
  - What happens to Endpoint objects when Pods are rescheduled?
  - Compare DNS-based vs environment variable service discovery methods.
- ReplicaSets and Deployments
  - Why would you manually create a ReplicaSet instead of using a Deployment?
  - What happens during a Deployment's rolling update process?
  - How can you rollback a failed Deployment using kubectl?
- DaemonSets and Jobs
  - When would you choose a DaemonSet over a Deployment?
  - What kubectl command shows completion status of Job objects?
  - How do you configure a Job to automatically clean up completed Pods?
- ConfigMaps and Secrets
  - Demonstrate creating a ConfigMap from a configuration file.
  - Why are Secrets base64-encoded but not encrypted by default?
  - How would you securely mount a Secret as a volume in a Pod?
- Persistent Data Management
  - Compare PersistentVolumes vs hostPath volumes for stateful applications.
  - What StorageClass parameters would you set for high-performance SSD storage?
  - How does StatefulSet guarantee persistent storage ordering during scaling?
  - What troubleshooting steps would you take if a Pod can't mount its PersistentVolumeClaim?
  - Explain how ReadWriteMany volumes enable shared access across multiple Pods.

Chapter 7
Kubernetes Resource Management for Cloud-Native and Edge Applications

Abstract This chapter examines Kubernetes resource management strategies tailored for both cloud-native and edge applications. It begins by defining key parameters for allocation, including CPU and memory requests and limits, pod density estimation, network bandwidth requirements, and disk I/O throughput. A cluster-wide resource allocation strategy is then outlined, using objective functions and constraints to balance efficiency with system reliability. Performance evaluation metrics such as CPU and memory utilization, network throughput, disk I/O, and energy efficiency are presented to assess allocation effectiveness. The chapter further explores scaling strategies, contrasting reactive autoscaling, which adjusts resources based on current load, with proactive autoscaling, which leverages forecasting techniques to anticipate demand. A comparative performance analysis demonstrates the trade-offs between responsiveness and predictive accuracy. By integrating quantitative metrics with autoscaling strategies, this chapter provides a comprehensive framework for optimizing Kubernetes resource management across diverse application environments.

Keywords Resource management · Kubernetes · CPU requests · Memory limits · Pod density · Autoscaling · Reactive scaling · Proactive scaling · Cloud-native Applications

Kubernetes resource management plays a vital role in optimizing performance, resource management, Load balancing, scalability, and cost efficiency for cloud-native and edge applications [95]. It ensures that workloads receive adequate computing, memory, storage, and network resources while maintaining system stability. Efficient resource allocation prevents both underutilization and over-provisioning, striking a balance between performance and cost. A fundamental aspect of Kubernetes resource management is defining key parameters for resource allocation. CPU and memory requests and limits determine the compute and memory resources allocated to each pod, ensuring workload stability. Pod density estimation helps in optimizing resource utilization by calculating the number of pods that can be scheduled within available resources. Additionally, network

B. Kumar et al., *Modern Kubernetes: From Core Concepts to Intelligent Autoscaling for Cloud Applications*, Studies in Autonomic, Data-driven and Industrial Computing, https://doi.org/10.1007/978-3-032-12972-7_7

bandwidth estimation ensures smooth data transfer, while disk I/O throughput calculation helps manage storage operations efficiently. Beyond individual pod management, cluster-wide resource allocation strategies are crucial for optimizing overall system performance. Kubernetes enforces resource constraints at the cluster level to minimize resource wastage while maintaining performance. These constraints include CPU and memory requests, disk storage, and network bandwidth allocation, ensuring fair resource distribution across multiple workloads. To assess the effectiveness of resource allocation, performance metrics play a key role. Metrics such as CPU utilization efficiency, memory utilization efficiency, network throughput efficiency, disk I/O efficiency, and energy consumption efficiency help in evaluating how well resources are used. Monitoring these metrics allows Kubernetes administrators to fine-tune autoscaling strategies, ensuring efficient workload execution. Autoscaling in Kubernetes can be categorized into reactive and proactive approaches. Reactive autoscaling dynamically adjusts resources based on real-time workload demands, using predefined thresholds such as CPU and memory usage. In contrast, proactive autoscaling leverages forecasting models to anticipate future resource needs and allocate resources accordingly. Proactive scaling helps prevent performance degradation due to sudden workload spikes, ensuring smoother operation. Finally, Kubernetes enables efficient resource management for cloud-native and edge applications by balancing real-time and predictive autoscaling approaches. By leveraging key performance metrics and optimized allocation strategies, organizations can achieve cost-effective, high-performance workload execution. The combination of reactive and proactive autoscaling enhances system reliability, ensuring seamless scalability in dynamic cloud and edge environments.

7.1 Defining Key Parameters for Resource Allocation

Resource allocation in Kubernetes is a crucial aspect of managing cloud-native and edge applications efficiently. It ensures that workloads receive the necessary compute, memory, storage, and network resources while maintaining scalability, stability, and cost-efficiency. Kubernetes allows users to define resource requests and limits for CPU, memory, network bandwidth, and disk I/O, preventing resource contention and optimizing workload execution. This section discusses key parameters for resource allocation in Kubernetes, including their definitions, calculations, and impact on system performance [96].

7.1.1 CPU Request and CPU Limit Calculation

The Central Processing Unit (CPU) is a vital resource in Kubernetes, determining the amount of computational power available to pods. Kubernetes allows users to specify CPU requests and limits to control resource consumption.

- CPU Request: The minimum guaranteed amount of CPU allocated to a pod. The Kubernetes scheduler ensures that a pod with a specific CPU request is assigned to a node with sufficient resources.
- CPU Limit: The maximum CPU a pod can consume before being throttled. If a pod exceeds this limit, Kubernetes restricts its CPU usage to ensure fair resource distribution.
- Formula: The CPU request for a pod can be estimated as:

$$CPU\_Request = Baseline\_CPU\_Usage + Expected\_CPU\_Spikes \tag{7.1}$$

  where:

  - **Baseline CPU Usage:** The average CPU required under normal conditions.
  - **Expected CPU Spikes:** Additional CPU demand during peak usage or unexpected workload surges.

- Impact:

  - A higher CPU request ensures stability, preventing performance degradation due to insufficient resources. However, excessive allocation may lead to over-provisioning, increasing costs and reducing cluster efficiency.
  - A CPU limit prevents excessive usage, ensuring fair resource distribution across pods. However, an overly restrictive CPU limit may cause performance throttling, affecting response times.

Proper tuning of CPU requests and limits is essential for maintaining application performance while optimizing cloud costs.

7.1.2 Memory Request and Memory Limit Calculation

Memory is a non-compressible resource in Kubernetes. Memory in Kubernetes is a critical resource that cannot be compressed or throttled like the CPU [96]. This means that once a pod is allocated memory, it cannot be reclaimed until the pod releases it. Unlike CPU, where excess usage can be controlled through throttling, memory must be carefully managed to prevent system failures. If a pod exceeds its memory allocation, Kubernetes may either terminate the pod (triggering an Out-Of-Memory (OOM) error) or evict it to free up resources and maintain cluster stability. Proper memory management ensures that applications run efficiently without unexpected failures.

- Memory Request: A memory request defines the minimum amount of memory guaranteed to a pod. Kubernetes uses this value to schedule pods onto nodes that have sufficient available memory. Setting an appropriate memory request ensures that the pod receives the necessary resources to function correctly under normal conditions.

- Memory Limit: A memory limit is the maximum amount of memory a pod can consume before Kubernetes intervenes. If a pod attempts to use more than its specified memory limit, Kubernetes will forcibly terminate it with an OOM error to prevent excessive resource consumption. This safeguard ensures that no single pod monopolizes available memory, maintaining a fair distribution of resources across all workloads.
- Formula: Kubernetes calculates the required memory request using the following equation:

$$Memory\_Request = Avg\_Memory\_Usage + Memory\_Overhead \tag{7.2}$$

  where:

  - **Avg. Memory Usage:** Represents the average memory consumption of an application under normal operational conditions. This value is determined by analyzing past memory usage patterns.
  - **Memory Overhead:** An additional buffer memory allocated to handle sudden spikes in memory consumption due to unpredictable workload surges.

  By incorporating memory overhead into the request, Kubernetes ensures that applications can accommodate transient workload increases without exceeding their limits.
- Impact:
  - Preventing Pod Eviction: Kubernetes schedules pods based on resource availability. If a pod's memory request is set too low, the scheduler might place it on a node with insufficient memory, leading to frequent evictions. Properly setting memory requests prevents unnecessary pod disruptions.
  - Avoiding Out-Of-Memory (OOM) Errors: When a pod exceeds its assigned memory limit, it is forcibly terminated with an OOM error. Setting appropriate memory limits helps prevent individual pods from consuming excessive memory, ensuring stable cluster operation.
  - Ensuring Efficient Resource Utilization: Allocating just the right amount of memory ensures optimal use of Kubernetes cluster resources. Over-provisioning can lead to wasted memory, reducing overall cluster efficiency, while under-provisioning may cause application failures.

- Practical Kubernetes Commands for Configuring Memory Requests and Limits In Kubernetes, memory requests and limits are defined in pod configuration files using YAML. Below is an example configuration for setting memory requests and limits:

```
apiVersion: v1
kind: Pod
metadata:
  name: example-pod
spec:
  containers:
```

```
  - name: example-container
    image: my-app-image
    resources:
      requests:
        memory: "256Mi"    # Minimum guaranteed memory
      limits:
        memory: "512Mi"    # Maximum memory allowed before
            termination
```

Explanation of the YAML Configuration: firstly is requested.memory: Defines the minimum memory allocation for the pod (256 MiB in this case). Kubernetes ensures that this amount is available before scheduling the pod. and second is limits.memory: Specifies the maximum memory a pod can use (512 MiB here). If exceeded, Kubernetes terminates the pod to prevent excessive memory consumption.

- Verifying Memory Requests and Limits in Kubernetes Once the pod is deployed, you can check its memory allocation details using the following command:

```
kubectl describe pod example-pod
```

This command provides information on the pod's assigned memory request and limit, along with any observed violations.

Memory management in Kubernetes plays a crucial role in ensuring application reliability and efficient resource utilization. By carefully defining memory requests and limits, administrators can prevent pod evictions, avoid OOM errors, and ensure fair resource distribution within the cluster. Implementing best practices and continuous monitoring further optimizes resource allocation, making Kubernetes-based cloud-native and edge applications more robust and scalable.

7.1.3 Pod Density Estimation

Pod density in Kubernetes refers to the number of pods that can run on a single node, considering the available CPU and memory resources. This metric is crucial for optimizing resource allocation and ensuring that a Kubernetes cluster efficiently utilizes its computational capacity without overloading nodes [97]. A high pod density means more efficient resource utilization, but it must be balanced to prevent performance degradation. If too many pods are scheduled on a node, they may compete for resources, causing throttling, memory exhaustion, and degraded performance. Proper pod density estimation helps in maintaining an optimal balance between resource allocation and application performance. **Formula:** The pod density is calculated using the following equation:

$$Pod\_Density = \frac{Total\_Available\_CPU/Memory}{Pod\_Request} \tag{7.3}$$

where: Total Available CPU/Memory: The total CPU and memory available on a Kubernetes node and Pod Request: The amount of CPU and memory a single pod requires. This formula helps determine the number of pods that can be scheduled on a node while ensuring sufficient resource allocation for each pod.

- Impact of Pod Density on Kubernetes Performance:
  - Efficient Resource Utilization: A higher pod density ensures that nodes are fully utilized, preventing idle resources and maximizing cost-effectiveness. If pod density is too low, it leads to inefficient cluster utilization, with some nodes remaining underused.
  - Risk of Performance Degradation: If too many pods are scheduled on a node, they may start competing for CPU and memory resources. This can lead to throttling, where some pods get fewer resources than required, impacting application performance. In extreme cases, pods may even get evicted if the node runs out of memory.
  - Avoiding Overloading and Throttling: Setting an appropriate pod density helps prevent Kubernetes from overloading a node. Overloading can result in higher latency, pod restarts due to resource contention, and even node crashes in severe cases.
- Practical Kubernetes Commands for Pod Density Management: In Kubernetes, pod density can be controlled by setting CPU and memory requests in the pod specification file. Below is an example YAML configuration:

```yaml
apiVersion: v1
kind: Pod
metadata:
  name: example-pod
spec:
  containers:
  - name: example-container
    image: my-app-image
    resources:
      requests:
        cpu: "250m"   # Minimum guaranteed CPU (250 millicores)
        memory: "512Mi"   # Minimum guaranteed memory (512 MiB)
```

  This configuration helps Kubernetes schedule the pod efficiently by ensuring that it is assigned appropriate CPU and memory resources.
- Verifying Pod Density in Kubernetes: To check the number of pods running on a node, use the following command:

```
kubectl get pods --all-namespaces -o wide
```

To check available resources on a node:

```
kubectl describe node <node-name>
```

This command provides information about CPU and memory usage, helping administrators determine whether pod density is optimal or if adjustments are needed.

Pod density estimation is an essential aspect of Kubernetes resource management. By calculating pod density using the available CPU and memory resources, Kubernetes administrators can optimize cluster performance while avoiding resource contention and overloading. Implementing best practices such as monitoring, autoscaling, and fine-tuning resource requests ensures an efficient and stable Kubernetes environment.

7.1.4 Network Bandwidth Estimation

Network bandwidth estimation in Kubernetes is crucial for ensuring efficient communication between pods, nodes, and external systems. It helps determine the required data transmission capacity per pod to prevent network congestion and performance degradation. In cloud-native and edge computing environments, where applications rely heavily on microservices and distributed architectures, proper bandwidth estimation ensures seamless data exchange [98–100].

- Key Considerations for Bandwidth Estimation Network bandwidth estimation involves evaluating the following parameters:
  - Pod-to-Pod Communication: The amount of data exchanged between pods within the cluster and it Affects latency-sensitive applications such as streaming and real-time processing.
  - Ingress and Egress Traffic: Incoming traffic (ingress) and outgoing traffic (egress) influence the bandwidth needed for external communication and are essential for applications that communicate with APIs, databases, or cloud services.
  - Service Mesh Overhead: Service meshes like Istio introduce additional network overhead due to encryption, retries, and logging and Must be factored into bandwidth estimation for microservices-heavy applications.
  - Edge Computing Constraints: In edge environments, limited bandwidth availability can impact application performance and Bandwidth estimation helps optimize data transmission between edge nodes and cloud data centers.
- Impact of Bandwidth Estimation on Kubernetes Performance
  - Preventing Network Bottlenecks and Latency: Insufficient bandwidth allocation can lead to network congestion, causing increased latency and degraded application performance. This is especially critical in applications that require

real-time data processing, such as video streaming, online gaming, and IoT systems.
  - Ensuring Smooth Communication Between Microservices: Cloud-native applications rely on microservices that communicate frequently. Proper bandwidth estimation ensures that microservices can exchange data efficiently without delays or dropped packets, improving the overall responsiveness of the application.
  - Optimizing Load Balancing and Traffic Distribution: Bandwidth estimation helps Kubernetes balance traffic across multiple pods and nodes. It allows efficient routing of requests to avoid overloading a particular node or network link.
  - Reducing Costs in Cloud and Edge Environments: Cloud providers charge for data transfer, making bandwidth estimation essential for cost control. Over-provisioning can lead to unnecessary expenses, while under-provisioning can cause performance issues.

- Kubernetes Tools and Commands for Bandwidth Estimation
  - Measuring Network Bandwidth Usage: Use kubectl top pod to check network usage metrics:

```
kubectl top pod --containers
```

    This command displays real-time CPU, memory, and network usage for each container.
  - Monitoring Network Traffic with Prometheus and Grafana: Install Prometheus and Grafana to monitor network bandwidth over time and Use the cAdvisor or Kube-state-metrics exporter to collect network traffic metrics.
  - Simulating Network Traffic Using lperf: Deploy Iperf to measure available bandwidth between two pods:

```
kubectl run iperf-server --image=networkstatic/iperf3 --
    port=5201
kubectl expose pod iperf-server --type=ClusterIP --port
    =5201
Run a client pod to test bandwidth:
```

    Run a client pod to test bandwidth:

```
kubectl run iperf-client --image=
    networkstatic/iperf3 --command -- iperf3
     -c iperf-server
```

    This helps determine actual bandwidth availability within the cluster.

Network bandwidth estimation is essential for ensuring smooth communication between Kubernetes pods, services, and external systems. By properly estimating and allocating bandwidth, organizations can prevent latency issues, optimize microservice interactions, and reduce cloud costs. Tools like Prometheus, Grafana, and Iperf help monitor and measure network performance, while Kubernetes CNI plugins enable fine-grained control over bandwidth usage.

7.1.5 Disk I/O Throughput Calculation

Disk I/O (Input/Output) refers to the read and write operations performed by a pod when interacting with persistent storage. It plays a crucial role in determining the efficiency of database transactions, file system operations, and overall storage performance in Kubernetes. Disk I/O throughput is a key metric for evaluating the speed at which data is read from and written to a disk over a given time period. Formula for Disk I/O Throughput Calculation [101].

$$Disk\_Throughput = \frac{Total\_Read + Write\_Operations}{Time\_Interval} \tag{7.4}$$

Where: **Total_Read:** The number of bytes read from storage over a given period, **Write_Operations:** The number of bytes written to storage over a given period and **Time_Interval:** The duration (in seconds) over which disk activity is measured.

This formula helps determine the overall efficiency of disk operations by measuring the rate of data transfer per second.

- Impact:
  - High Disk Throughput Ensures Smooth Database Operations: Applications that require frequent disk access, such as databases, caching systems, and logging services, depend on high disk throughput for optimal performance. Slow disk reads/writes can lead to query delays, sluggish application responses, and degraded user experience.
  - Poor Disk Performance Leads to Storage Bottlenecks: When disk I/O throughput is low, storage bottlenecks occur, causing increased latency in data retrieval and processing. This can impact Kubernetes workloads in various ways:

    Slow database queries: Databases like MySQL, PostgreSQL, and MongoDB rely on fast disk operations for efficient data access.
    Delayed log processing: Logging services such as ELK Stack (Elasticsearch, Logstash, Kibana) generate large volumes of disk writes, requiring high I/O throughput.
    Increased pod restart times: Applications relying on persistent storage may take longer to recover from failures.

- Monitoring Disk I/O in Kubernetes: To ensure optimal disk performance, Kubernetes administrators can monitor disk throughput using various tools and commands.

- Checking Disk I/O with kubectl top pod: Kubernetes provides basic resource usage statistics, but does not directly show disk I/O metrics. However, you can use the following command to check CPU and memory usage:

```
kubectl top pod --containers
```

  For detailed disk I/O metrics, additional tools are required.
- Using iostat to Measure Disk Performance: The iostat command provides real-time disk I/O statistics. Run it inside a Kubernetes node:

```
iostat -dx 1
```

  This displays detailed disk read/write throughput statistics.
- Using Prometheus and Grafana for Disk Monitoring: Prometheus Node Exporter collects disk I/O metrics, including read/write rates. Grafana Dashboards visualize disk performance over time. Example PromQL query to monitor disk read and write throughput:

```
rate(node_disk_read_bytes_total[5m])
rate(node_disk_written_bytes_total[5m])
```

- Measuring Disk I/O with fio (Flexible I/O Tester): fio is a powerful tool for benchmarking disk performance. Install and run it inside a Kubernetes pod:

```
fio --name=test --rw=write --bs=4k --size=1G --numjobs=1 --
    time_based --runtime=60
```

  This command simulates disk write operations, measuring throughput and latency.

Kubernetes resource allocation plays a pivotal role in ensuring stable, efficient, and cost-effective workload execution. By carefully configuring CPU, memory, network bandwidth, and disk I/O parameters, Kubernetes administrators can:

- Prevent resource contention and ensure fair distribution.
- Optimize application performance and avoid CPU throttling or memory crashes.
- Enhance scalability by dynamically adjusting resources.
- Reduce cloud costs by preventing over-provisioning.

As workloads grow in complexity, automated resource scaling using Kubernetes Horizontal Pod Autoscaler (HPA) and Vertical Pod Autoscaler (VPA) can further optimize performance and resilience.

7.2 Cluster-Wise Resource Allocation Strategy

Efficient resource allocation in a Kubernetes cluster is essential for optimizing performance and reducing wastage. This strategy ensures that computing resources, such as CPU and memory, are distributed effectively among pods while meeting application demands.

7.2.1 *Objective Function for Resource Allocation*

The primary objective of resource allocation is to minimize resource wastage while ensuring that each pod receives the necessary resources to operate optimally. Kubernetes dynamically schedules and allocates resources based on pod requests and available node capacity. The formula for Resource Optimization:

$$\min \sum_{i=1}^{N} (Allocated\_Resources - Utilized\_Resources)^2 \tag{7.5}$$

Where: Allocated_Resources: The total CPU/memory allocated to a pod, Utilized_Resources: The actual CPU/memory consumed by the pod and N: The total number of pods in the cluster. This objective function ensures that the allocated resources closely match the actual utilization, reducing inefficiencies such as underutilization or over-provisioning.

- Impact of Optimized Resource Allocation:
  - Cost Efficiency: By optimizing resource allocation through Kubernetes autoscaling, organizations can significantly reduce cloud infrastructure costs. Autoscaling prevents unnecessary over-provisioning by dynamically adjusting CPU, memory, and storage resources based on real-time demand. This ensures that businesses only pay for what they use, avoiding wasted capacity. For example, during periods of low traffic, the system scales down pod deployments, freeing up resources and minimizing expenses. Additionally, proactive autoscaling techniques, such as predictive scaling using machine learning, further enhance cost savings by anticipating workload fluctuations before they occur, allowing for smoother and more efficient resource adjustments [102].
  - Improved Performance: Autoscaling ensures that workloads consistently receive the resources they need, preventing performance bottlenecks and maintaining application responsiveness. When demand spikes, Kubernetes automatically provisions additional pods to distribute the load, reducing latency and avoiding service degradation. Conversely, during low-traffic periods, it scales down to prevent resource contention. Efficient memory and CPU utilization, guided by predefined requests and limits, further stabilize per-

formance by preventing individual pods from monopolizing node resources. This dynamic resource management is particularly critical in latency-sensitive environments, such as edge computing, where delays can directly impact user experience.
- Scalability: One of the most powerful features of Kubernetes autoscaling is its ability to dynamically adjust cluster capacity in response to changing workloads. Horizontal Pod Autoscaling (HPA) and Cluster Autoscaler work together to ensure that applications can seamlessly scale out (adding more pods) or scale up (allocating more resources) as demand increases—and scale back down when demand subsides. This elasticity is essential for handling unpredictable traffic patterns, such as sudden surges during sales events or seasonal peaks. In edge computing environments, where resources may be constrained, autoscaling ensures that workloads are efficiently distributed across available nodes, maintaining reliability without manual intervention. This flexibility makes Kubernetes an ideal platform for modern cloud-native and distributed applications.

7.2.2 Constraints for Kubernetes Cluster Allocation

To prevent overloading and ensure that Kubernetes effectively schedules pods, resource allocation must adhere to constraints on CPU and memory.

- CPU Constraints:

$$\sum Pod\_CPU\_Requests \leq Total\_Node\_Capacity \tag{7.6}$$

  Ensures that the total CPU requests of all running pods do not exceed the total CPU capacity of the node and prevents CPU starvation, which could degrade application performance.
- Memory Constraints:

$$\sum Pod\_Memory\_Requests \leq Total\_Node\_Memory \tag{7.7}$$

  Ensures that memory requests do not exceed the node's available memory and Prevents Out-of-Memory (OOM) errors, which can cause pod evictions and failures.

Overall, Kubernetes enforces CPU and memory constraints to prevent node overloading, ensuring that workloads are scheduled only within the node's available capacity. Without these limits, a single resource-intensive pod could consume all available CPU or memory, starving other applications and destabilizing the entire node. By defining appropriate requests and limits, administrators maintain a balanced distribution of resources, allowing the cluster to operate smoothly even under fluctuating demand. These constraints also enhance system stability by reducing the risk of crashes and performance degradation. When a pod exceeds its memory limit, Kubernetes terminates it to protect the node from becoming unresponsive. Similarly, CPU throttling prevents runaway processes from monopolizing compute power. This proactive management minimizes downtime and ensures consistent application performance, particularly in production environments where reliability is critical. Additionally, CPU and memory constraints promote fair resource distribution among multiple pods. By setting explicit resource requests, Kubernetes ensures that each workload receives its guaranteed share, preventing resource starvation. This is especially important in multi-tenant clusters, where different teams or applications share the same infrastructure. Fair scheduling improves overall efficiency, allowing more workloads to run reliably on fewer nodes while reducing infrastructure costs.

By adhering to these constraints, Kubernetes optimizes resource allocation, maximizing both efficiency and system reliability. Properly configured limits enable the cluster to scale intelligently, handle spikes in demand, and maintain high availability—key requirements for modern cloud-native and edge computing environments.

7.3 Performance Metrics for Evaluating Resource Allocation

Evaluating resource allocation efficiency in Kubernetes is essential to ensure optimal performance, reduce waste, and maintain system stability. Several key performance metrics help assess how well resources like CPU, memory, network bandwidth, disk I/O, and energy are utilized [103].

7.3.1 CPU Utilization Efficiency

CPU utilization efficiency measures how effectively allocated CPU resources are used. If a pod is allocated more CPU than required, it leads to resource wastage, while insufficient allocation can cause performance bottlenecks. The formula calculates CPU efficiency as the ratio of actual CPU usage to allocated CPU, expressed as a percentage, ensuring optimal utilization.

$$CPU\_Efficiency = \left(\frac{Actual\_CPU\_Usage}{Allocated\_CPU}\right) \times 100\% \tag{7.8}$$

Impact:

- High CPU efficiency indicates optimal resource usage and better performance.
- Low CPU efficiency suggests underutilization, leading to inefficient resource allocation.

7.3.2 Memory Utilization Efficiency

Memory utilization efficiency ensures that allocated memory is effectively used by pods. Over-provisioning results in resource waste, while under-provisioning can cause Out-Of-Memory (OOM) errors. The formula computes memory efficiency by comparing actual memory usage to allocated memory, expressed as a percentage, ensuring optimal resource allocation [].

$$Memory\_Efficiency = \left(\frac{Actual\_Memory\_Used}{Allocated\_Memory}\right) \times 100\% \tag{7.9}$$

Impact:

- Low efficiency indicates excessive memory allocation, wasting resources.
- High efficiency (>90%) increases the risk of OOM errors if memory demand spikes unexpectedly.

7.3.3 Network Throughput Efficiency

This metric evaluates how efficiently network resources are utilized for data transmission. Inefficient network allocation can cause congestion and increased latency. The formula calculates network efficiency by measuring the proportion of utilized network bandwidth based on the total data transmitted and received. A higher value indicates better bandwidth utilization [98].

$$Network\_Efficiency = \frac{Total\_Data\_Sent + Total\_Data\_Received}{Total\_Network\_Bandwidth} \tag{7.10}$$

Impact:

- High network efficiency ensures smooth communication between microservices.
- Poor efficiency can lead to network bottlenecks, reducing system performance.

7.3.4 Disk I/O Efficiency

Disk I/O efficiency measures the effectiveness of read/write operations within a given storage bandwidth. Poor disk utilization can slow down application performance, especially for data-intensive workloads. The formula calculates disk efficiency by evaluating the proportion of disk I/O operations (reads and writes) relative to the maximum allowable disk throughput. A higher value signifies optimal disk utilization, whereas a lower value may indicate resource underutilization [101].

$$Disk\_Efficiency = \frac{Total\_Read + Total\_Write}{Disk\_I/O\_Limit} \tag{7.11}$$

Impact:

- High disk efficiency ensures smooth storage operations.
- Poor disk performance results in slow read/write speeds, affecting overall application responsiveness.

7.3.5 Energy Consumption Efficiency

This metric determines how effectively energy is utilized for executing tasks within a Kubernetes cluster. Reducing energy consumption while maintaining performance is critical for sustainable cloud computing. The formula for energy efficiency measures the number of tasks completed per unit of energy consumed. A higher value indicates better resource optimization and lower energy wastage, which is crucial for sustainable cloud and edge computing environments.

$$Energy\_Efficiency = \frac{Total\_Tasks\_Completed}{Total\_Energy\_Used} \tag{7.12}$$

Impact:

- Efficient energy utilization reduces operational costs and environmental impact.
- Poor energy efficiency leads to unnecessary power consumption, increasing costs.

These metrics provide a structured way to assess resource allocation, helping Kubernetes administrators optimize their clusters for performance, cost savings, and sustainability.

7.4 Reactive and Proactive Autoscaling Algorithms

Autoscaling in cloud computing is a mechanism that automatically adjusts the number of running instances based on demand. There are two primary approaches: **Reactive Autoscaling**, which responds to real-time workload changes, and **Proactive Autoscaling**, which predicts future demand to adjust resources in advance [104].

7.4.1 *Reactive Autoscaling*

Reactive autoscaling dynamically adjusts resources based on real-time demand, responding to usage spikes and reducing excess capacity when demand decreases.

- Mathematical Model The decision to scale up or down is based on a threshold condition:

$$Scaling\_Trigger = \frac{Current\_Usage}{Scaling\_Threshold} \tag{7.13}$$

  where:

  - *Current_Usage* represents CPU or memory utilization.
  - *Scaling_Threshold* is the predefined limit that triggers scaling.

- Algorithm for Reactive Autoscaling

Algorithm 1 Reactive autoscaling algorithm

Initialize CPU_threshold, Memory_threshold, Min_pods, Max_pods, Scaling_factor
while Application is running **do**
  Monitor resource usage
  if (CPU_usage > CPU_threshold OR Memory_usage > Memory_threshold) **then**
    Compute new pod count:

$$New\_pod\_count = \min(Max\_pods, Current\_pod\_count + Scaling\_factor)$$

    Deploy additional pods
  else if (CPU_usage < 0.7 × CPU_threshold AND Memory_usage < 0.7 × Memory_threshold) **then**
    Compute new pod count:

$$New\_pod\_count = \max(Min\_pods, Current\_pod\_count - Scaling\_factor)$$

    Terminate excess pods
  end if
  Wait for next monitoring interval
end while

7.4.2 *Proactive Autoscaling*

Proactive autoscaling uses machine learning and historical data to predict future resource demand, adjusting resource allocation before a surge occurs [105].

- Mathematical Model Predictive models such as ARIMA, LSTM, or Reinforcement Learning are used. A common forecasting equation:

$$Predicted\_Usage = \sum_{i=1}^{n}(w_i \times Past\_Utilization_i) \tag{7.14}$$

  where:

  - *Past_Utilization* represents historical CPU/memory usage.
  - w_i is the weight assigned to past data points.

- Algorithm for Proactive Autoscaling

Algorithm 2 Proactive autoscaling algorithm

Initialize system parameters: CPU_threshold, Memory_threshold, Min_pods, Max_pods, Prediction_model
Train Prediction_model using historical CPU and Memory data
while Application is running **do**
  Forecast future resource utilization:

$$Predicted\_CPU = Prediction\_model.forecast(Forecasting\_window)$$

$$Predicted\_Memory = Prediction\_model.forecast(Forecasting\_window)$$

  if (Predicted_CPU > CPU_threshold OR Predicted_Memory > Memory_threshold) **then**
    Compute required pod count:

$$Required\_pods = \lceil \frac{Predicted\_CPU}{CPU\_threshold} \rceil$$

    Scale up resources
  else if (Predicted_CPU < 0.7 × CPU_threshold AND Predicted_Memory < 0.7 × Memory_threshold) **then**
    Scale down resources
  end if
  Update Prediction_model with new data
end while

7.4.3 Comparative Performance Evaluation

Comparative Performance Evaluation analyzes the efficiency of reactive and proactive autoscaling strategies in dynamically managing Kubernetes resources, balancing responsiveness and predictive stability.

Table 7.1 Performance comparison of autoscaling strategies

Metric	Reactive autoscaling	Proactive autoscaling
Response Time	Slower (reactive)	Faster (predictive)
CPU Efficiency	Moderate	High
Pod Density Management	Fluctuating pods	Stable pod provisioning
Network Bandwidth	High spikes	Smooth bandwidth usage
Disk I/O Load	Higher	Lower
Energy Consumption	Higher (frequent scaling)	Lower (optimized scaling)

Reactive autoscaling is a simple and commonly used technique but suffers from delayed responses, inefficient resource usage, and higher energy consumption (Table 7.1). Proactive autoscaling, by predicting future resource needs, achieves better resource efficiency, stable network bandwidth, and optimized energy consumption. A hybrid approach combining both methods can further improve system efficiency.

7.5 Key Takeaways

- Defining Key Parameters for Resource Allocation
  - CPU Request and Limit Calculation: Determines the minimum and maximum CPU resources allocated to a pod to avoid over- or under-utilization.
  - Memory Request and Limit Calculation: Ensures efficient memory allocation, preventing out-of-memory (OOM) errors while optimizing resource consumption.
  - Pod Density Estimation: Defines how many pods can be efficiently deployed per node based on available resources.
  - Network Bandwidth Estimation: Assesses required data transfer rates to prevent network bottlenecks in cloud and edge environments.
  - Disk I/O Throughput Calculation: Measures read/write operations to optimize storage performance and avoid latency.
- Cluster-Wise Resource Allocation Strategy
  - Objective Function for Resource Allocation: Optimizes resource usage based on workload demands while ensuring scalability and performance.
  - Constraints for Kubernetes Cluster Allocation: Includes CPU/memory limits, QoS policies, affinity/anti-affinity rules, and scheduling constraints to maximize efficiency.
- Performance Metrics for Evaluating Resource Allocation
  - CPU Utilization Efficiency: Measures how effectively CPU resources are utilized to prevent under-provisioning or wastage.

  - Memory Utilization Efficiency: Ensures memory is optimally allocated without causing system slowdowns or excessive consumption.
  - Network Throughput Efficiency: Evaluates data transfer efficiency across pods, nodes, and external services.
  - Disk I/O Efficiency: Ensures storage read/write operations are optimized for application performance.
  - Energy Consumption Efficiency: Reduces power usage by optimizing resource allocation and scaling mechanisms.

- Reactive and Proactive Autoscaling Algorithms
  - Reactive Autoscaling: Adjusts resources based on current demand, responding to CPU, memory, or network usage spikes.
  - Proactive Autoscaling: Uses predictive models (e.g., ARIMA, LSTM) to forecast future demand and allocate resources in advance.

- Comparative Performance Evaluation
  - Reactive Scaling: Fast response to demand changes but can lead to delays and inefficient resource usage.
  - Proactive Scaling: Anticipates future needs, improving resource efficiency and stability.
  - Hybrid Approach: Combining both methods can enhance overall performance and minimize costs.

7.6 Summary

Efficient resource management in Kubernetes is essential for ensuring the performance and scalability of cloud-native and edge applications. This involves defining **CPU and memory limits**, optimizing **pod density**, and accurately estimating **network bandwidth** and **disk I/O requirements**. Proper resource allocation not only prevents bottlenecks but also improves overall system efficiency. To achieve optimal resource usage, **cluster-wide allocation strategies** are employed, considering constraints such as **scheduling policies**, **CPU/memory availability**, and **workload balancing**. These strategies help maintain system stability while ensuring that nodes and pods operate within defined limits. Performance evaluation plays a critical role in assessing resource allocation effectiveness. Key metrics include **CPU utilization efficiency**, **memory utilization efficiency**, **network throughput efficiency**, **disk I/O efficiency**, and **energy consumption efficiency**. Monitoring these parameters provides insights into system performance, helping identify areas for optimization. Furthermore, **autoscaling strategies** dynamically adjust resources to match workload demands. **Reactive autoscaling** responds to real-time fluctuations based on predefined thresholds, ensuring immediate scaling but potentially struggling with sudden spikes. In contrast, **proactive autoscaling** leverages **historical data**

and predictive models to anticipate future demands, offering better stability and efficiency. While proactive methods require additional computational overhead, they generally lead to smoother performance and reduced latency. Overall, a combination of efficient resource allocation, performance monitoring, and strategic autoscaling mechanisms enables Kubernetes clusters to maintain optimal functionality, ensuring high availability and cost-effectiveness for cloud-native and edge-based applications.

7.7 Review Questions

- What is the difference between CPU **request** and **limit** in Kubernetes resource allocation?
- How does Kubernetes estimate **pod density**, and why is it important?
- What factors influence **network bandwidth estimation** in Kubernetes clusters?
- Why is **disk I/O throughput** an important consideration in resource allocation?
- What are the key constraints in **Kubernetes cluster allocation** strategies?
- How does **CPU utilization efficiency** impact overall cluster performance?
- What are the advantages and disadvantages of **reactive vs. proactive autoscaling**?
- How can **machine learning models** improve proactive autoscaling decisions?
- What are the trade-offs between **energy consumption efficiency** and resource availability?
- In what scenarios would a **hybrid autoscaling approach** be more effective than a purely reactive or proactive strategy?

Chapter 8
Reactive vs. Proactive Autoscaling in Kubernetes

Abstract This chapter explores the differences between reactive and proactive autoscaling approaches in Kubernetes and their role in optimizing cloud-native performance. It begins with an overview of Kubernetes autoscaling mechanisms and categories, contrasting reactive strategies, which respond to real-time workload fluctuations, with proactive strategies, which forecast future demand using predictive models. The integration of operator reconciliation with model-driven control systems is highlighted as a method for refining resource allocation. Performance evaluation is presented through real-world implementations, including Kubernetes–Docker environments and a case study on dynamic allocation in cloud computing. Key findings demonstrate the benefits of predictive autoscaling, including reduced latency, optimized cloud expenses, and improved response times across diverse workloads. The chapter concludes with a forward-looking perspective on advancing predictive autoscaling techniques for both cloud and edge computing, emphasizing the importance of intelligent, demand-aware orchestration in meeting evolving performance and cost requirements.

Keywords Autoscaling · Reactive scaling · Proactive scaling · Predictive models · Operator reconciliation · Cost optimization · Edge computing

Modern cloud-native applications operate in dynamic and unpredictable environments, where workload demand fluctuates due to varying user interactions, seasonal trends, and real-time processing requirements [106]. Efficient resource management is crucial to maintain optimal performance, minimize costs, and ensure high availability. Kubernetes, the leading container orchestration platform, offers built-in autoscaling mechanisms to address these challenges [107]. Autoscaling in Kubernetes refers to the ability to dynamically adjust computational resources based on changing workload demands [108]. By automatically scaling up or down pods, containers, and infrastructure components, Kubernetes ensures applications maintain their performance without excessive resource allocation. However, autoscaling strategies can be broadly classified into two categories: **Reactive Autoscaling**—Responds to workload variations after they occur, using real-time metrics such as

B. Kumar et al., *Modern Kubernetes: From Core Concepts to Intelligent Autoscaling for Cloud Applications*, Studies in Autonomic, Data-driven and Industrial Computing, https://doi.org/10.1007/978-3-032-12972-7_8

CPU and memory usage and **Proactive Autoscaling**—Uses predictive analytics and machine learning models to anticipate future workload trends and scale resources before demand surges. While reactive autoscaling is widely used in Kubernetes deployments, it often leads to delayed responses, resource over-provisioning, or under-provisioning, which can result in higher costs and performance bottlenecks. On the other hand, proactive autoscaling leverages forecasting techniques to enhance scalability decisions, ensuring optimal resource utilization. Modern cloud applications encounter unpredictable workload fluctuations, making autoscaling a crucial component for efficient resource management. Various real-world scenarios demonstrate its significance, such as **e-commerce** platforms experiencing traffic spikes during sales events, **streaming services** adjusting to peak-hour user activity, **microservices** architectures requiring dynamic resource distribution, and edge computing or IoT applications where workloads shift based on real-time events. **Without effective autoscaling**, these applications risk resource wastage due to over-provisioning, leading to unnecessary cloud expenses, or performance degradation caused by under-provisioning, resulting in service disruptions and poor user experiences. Additionally, manual scaling introduces higher operational costs, as it demands continuous monitoring and human intervention. To address these challenges, Kubernetes employs automated autoscaling mechanisms that dynamically allocate resources based on demand, ensuring optimal performance, cost efficiency, and seamless adaptability to varying workloads.

This chapter explores the limitations of traditional reactive autoscaling, introduces advanced proactive techniques, and discusses how machine learning-driven forecasting models can optimize Kubernetes autoscaling strategies [106, 109, 110]. It highlights the drawbacks of reactive autoscaling, such as latency issues and inefficient resource utilization, and explains how proactive autoscaling addresses these challenges by predicting workload fluctuations in advance [111]. Through case studies and performance evaluations, this chapter demonstrates how predictive scaling improves Kubernetes-based resource management.

8.1 Kubernetes Autoscaling Mechanisms

Kubernetes offers three primary autoscaling mechanisms—Horizontal Pod Autoscaler (HPA), Vertical Pod Autoscaler (VPA), and Cluster Autoscaler (CA)—to dynamically manage resources based on application workload demands. These mechanisms help ensure efficient utilization of compute resources, prevent performance degradation, and reduce operational costs by automating scaling decisions.

- Horizontal Pod Autoscaler (HPA): The Horizontal Pod Autoscaler (HPA) is one of the most widely used autoscaling mechanisms in Kubernetes. It dynamically adjusts the number of running pods based on real-time workload demands, ensuring that applications maintain performance under fluctuating traffic conditions.

  - How It Works: HPA monitors metrics such as CPU and memory usage and automatically increases or decreases the number of pods to maintain a predefined target utilization. Triggering Conditions: HPA operates based on threshold values set by users (e.g., scaling up when CPU utilization exceeds 70%).
  - Use Cases: It is particularly effective for stateless applications, such as web services and API servers, where adding or removing pods does not impact the application's overall state.
  - Limitations: HPA is reactive, meaning it adjusts scaling only after a workload change has already occurred, which can lead to delays in response time during sudden traffic spikes.

- Vertical Pod Autoscaler (VPA): The Vertical Pod Autoscaler (VPA) focuses on adjusting resource requests and limits within individual pods rather than changing the number of pods. It ensures that applications receive the necessary CPU and memory resources to perform optimally.

  - How It Works: VPA continuously analyzes resource utilization and dynamically updates CPU and memory allocations for each pod based on historical and real-time usage patterns. Prevention of Resource Starvation: It prevents issues such as Out of Memory (OOM) errors or CPU starvation, which can degrade application performance.
  - Use Cases: VPA is ideal for stateful applications, background processes, and workloads where increasing the number of pods is not feasible (e.g., databases, machine learning workloads).
  - Limitations: Unlike HPA, VPA cannot be used alongside it effectively because HPA scales out (adds pods), whereas VPA scales up (allocates more resources to existing pods). Additionally, updating pod resource allocations may require restarting the pod, leading to temporary downtime.

- Cluster Autoscaler (CA): The Cluster Autoscaler (CA) operates at the node level, dynamically adjusting the number of worker nodes in a Kubernetes cluster to ensure that there are enough compute resources available for scheduled pods.

  - How It Works: CA automatically provisions new virtual machines (VMs) or removes underutilized ones based on pending pod resource requests and overall cluster utilization. Integration with HPA and VPA: CA works in conjunction with HPA and VPA, ensuring that scaling at the pod level does not lead to resource bottlenecks at the cluster level.
  - Use Cases: It is essential for managing multi-tenant Kubernetes clusters, cloud environments where cost efficiency is a priority, and applications with fluctuating workloads that require infrastructure-level scaling.
  - Limitations: CA is highly dependent on cloud provider APIs, meaning its efficiency may vary based on the cloud environment (e.g., AWS, Azure, or Google Cloud). Additionally, scaling nodes can take more time compared to pod-level autoscaling, leading to potential delays in handling workload surges.

By leveraging HPA, VPA, and CA, Kubernetes ensures automated, efficient, and scalable resource allocation without the need for manual intervention. While HPA provides flexibility for scaling stateless workloads, VPA optimizes resource consumption for individual pods, and CA ensures that infrastructure-level resources are adequately provisioned. In practice, combining these autoscaling mechanisms strategically allows organizations to balance performance, cost, and reliability in cloud-native applications.

8.2 Categories of Autoscaling Strategies

Kubernetes autoscaling strategies can be broadly classified into two categories: Reactive Autoscaling and Proactive Autoscaling [108]. Each approach has distinct mechanisms, advantages, and challenges, impacting how effectively resources are managed in cloud-native environments. Autoscaling approaches can be categorized into reactive and proactive methods (see Fig. 8.1). Reactive autoscaling relies on threshold-based techniques and queuing theory to adjust resources based on real-time demand. In contrast, proactive autoscaling utilizes statistical approaches, reinforcement learning, fuzzy logic, machine learning, and time series analysis to predict and allocate resources in advance.

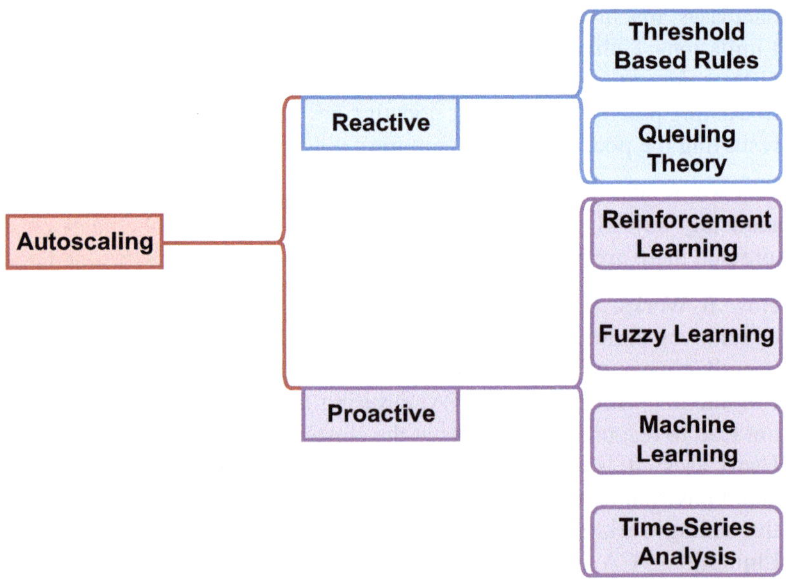

Fig. 8.1 Autoscaling approaches

8.2.1 Reactive Autoscaling

Reactive autoscaling responds to real-time system metrics, adjusting resources after workload changes occur [112]. It operates based on threshold values, making it suitable for handling predictable variations in application workloads. Threshold-based and queuing theory approaches are widely used in reactive autoscaling for managing cloud resources. **Threshold-based autoscaling** operates by defining upper and lower limits for system metrics such as CPU utilization, memory usage, or network traffic. When these thresholds are exceeded, the system triggers scaling actions, either adding or removing resources to maintain performance. While simple and easy to implement, this method may struggle with sudden workload spikes or gradual fluctuations. In contrast, **queuing theory-based autoscaling** models resource demand as a queueing system, analyzing arrival and service rates to determine optimal scaling decisions. By estimating queue length and system congestion, this approach dynamically adjusts resources to maintain service quality while minimizing over-provisioning. Though more adaptive than threshold-based methods, queuing theory requires accurate workload modeling and may introduce computational overhead in complex systems.

- Mechanism: Reactive autoscaling relies on historical data and predefined triggers (e.g., CPU utilization exceeding 80%). When resource usage surpasses a specified threshold, Kubernetes scales up by adding pods or increasing node capacity. If the workload decreases, the system scales down to optimize cost efficiency. It is primarily implemented using Horizontal Pod Autoscaler (HPA) and Cluster Autoscaler (CA) in Kubernetes.
- Advantages: **Simplicity**—Easy to implement and configure using built-in Kubernetes mechanisms. **Effective for predictable workloads**—Works well when resource demand follows a consistent pattern. **Integration with cloud providers**—Cloud-native autoscaling solutions, such as AWS Auto Scaling and Google Kubernetes Engine (GKE) Autoscaler, support reactive scaling natively.
- Challenges: **Latency Issues**—Scaling occurs only after performance degradation is detected, leading to temporary slowdowns. **Over-Provisioning**—To avoid delays, organizations may allocate excess resources, resulting in higher cloud costs. **Under-Provisioning**—Sudden traffic spikes may lead to resource shortages before scaling kicks in, affecting application performance.

8.2.2 Proactive Autoscaling

Proactive autoscaling is a resource management approach that predicts future workload demands and adjusts computing resources in advance to maintain optimal performance and efficiency [112]. It anticipates workload fluctuations by leveraging machine learning (ML) and predictive analytics to optimize resource allocation [113]. Unlike reactive autoscaling, which responds only after demand changes occur, proactive methods forecast trends and scale resources preemptively, ensuring system stability and efficiency. By analyzing historical workload patterns, these techniques predict future resource needs, enabling applications to sustain optimal performance without delays. This approach minimizes the risk of performance degradation during sudden traffic surges, ensuring a seamless user experience while optimizing infrastructure costs. One of the key benefits of proactive autoscaling is its ability to reduce latency by provisioning resources before demand spikes occur. This preemptive strategy ensures smooth application performance, preventing bottlenecks that could lead to system slowdowns or failures. Additionally, proactive autoscaling prevents over-provisioning, unlike reactive methods that may allocate excessive resources after detecting high usage. By optimizing resource allocation, proactive strategies help organizations reduce unnecessary cloud expenditures while maintaining high availability, ensuring applications remain responsive even under unpredictable workload variations. Despite its advantages, proactive autoscaling presents several challenges. Implementing predictive models requires complex integration with cloud infrastructure and continuous model training to maintain forecasting accuracy. The computational overhead of proactive autoscaling is also higher than reactive methods, as it demands additional processing power for real-time data analysis and prediction generation. Furthermore, the effectiveness of proactive autoscaling depends heavily on the quality of data used for training. Poor data quality or incorrect forecasting can lead to suboptimal resource allocation, causing inefficiencies in resource usage or potential performance issues.

The mechanism behind proactive autoscaling involves sophisticated predictive models that analyze past workload trends to anticipate future demands. This enables preemptive scaling, reducing the likelihood of latency, system failures, and resource wastage. Proactive autoscaling often integrates reinforcement learning techniques that continuously refine scaling decisions, deep learning models such as LSTM, Transformer, and InformerAutoScale for complex pattern recognition, and time-series forecasting algorithms like ARIMA, SARIMA, and Holt-Winters to enhance prediction accuracy [110, 114–117]. By combining these advanced methodologies, proactive autoscaling ensures efficient and intelligent resource provisioning, making it a highly effective solution for dynamic cloud environments.

1. Statistical Models: Use mathematical techniques to analyze historical data patterns and make predictions.
   - ARIMA (Autoregressive Integrated Moving Average): A widely used time-series forecasting model that captures historical trends to predict future resource needs.
   - SARIMA (Seasonal ARIMA): An extension of ARIMA that considers seasonal variations in workload patterns, making it useful for applications with recurring traffic surges.
   - Holt-Winters Exponential Smoothing: A forecasting method that accounts for trends and seasonality, helping in workload prediction for applications with periodic traffic variations.

2. Deep Learning Models: Utilize artificial neural networks to learn complex patterns from large datasets, often used for time-series forecasting.

   - LSTM (Long Short-Term Memory Networks): A type of recurrent neural network (RNN) that captures long-term dependencies in workload fluctuations, making it effective for autoscaling predictions [118].
   - Bi-LSTM (Bidirectional LSTM): Enhances LSTM by processing data in both forward and backward directions, leading to improved prediction accuracy.
   - GRU (Gated Recurrent Unit): A variant of LSTM that simplifies computations while maintaining similar predictive performance for time-series forecasting.
   - CNN-LSTM (Convolutional Neural Network + LSTM): Combines CNNs for feature extraction with LSTMs for sequential prediction, improving resource demand forecasting.
   - Transformer Model: Uses self-attention mechanisms to analyze complex workload dependencies, enabling more accurate scaling predictions.
   - InformerAutoScale: A Transformer-based model optimized for long-sequence time-series forecasting, reducing computational overhead while maintaining high accuracy.

3. Machine Learning Models: Apply algorithms that learn from data to make predictions and decisions without explicit programming.

   - Linear Regression: A simple model that predicts future workload based on past trends but may not handle complex patterns effectively.
   - Polynomial Regression: Extends linear regression by capturing non-linear relationships between workload variations and resource demands.
   - Random Forest Regression: An ensemble learning method that improves prediction accuracy by averaging multiple decision trees.
   - XGBoost (Extreme Gradient Boosting): A powerful gradient boosting algorithm that enhances predictive accuracy and efficiency, making it suitable for workload forecasting.

4. Reinforcement Learning Models: Optimize decision-making through trial and error, using rewards and penalties to improve resource allocation (e.g., Q-learning, Deep Q-Network).

   - Deep Q-Network (DQN): Uses Q-learning and neural networks to dynamically adjust scaling decisions based on observed system performance.
   - Proximal Policy Optimization (PPO): A reinforcement learning algorithm that optimizes autoscaling policies by balancing performance and cost.
   - Advantage Actor-Critic (A2C): Uses both an actor network (for decision-making) and a critic network (for evaluating performance) to improve autoscaling efficiency.

5. Fuzzy Logic Models: Utilize rule-based systems to handle uncertainty and imprecise data, making decisions based on linguistic variables rather than strict numerical thresholds.

- Mamdani Fuzzy Inference System: A widely used fuzzy logic approach that applies human-like reasoning with if-then rules to manage autoscaling decisions based on workload conditions.
- Sugeno Fuzzy Model: A computationally efficient fuzzy inference method that generates precise scaling decisions by mapping inputs to outputs using linear equations.
- Adaptive Neuro-Fuzzy Inference System (ANFIS): Combines neural networks with fuzzy logic to learn and optimize scaling rules dynamically based on historical workload patterns.
- Type-2 Fuzzy Logic System: Enhances traditional fuzzy logic by incorporating uncertainty modeling, making it suitable for handling unpredictable cloud workloads with varying demand.

6. Hybrid and Ensemble Models: Combine multiple models to enhance prediction accuracy and robustness, leveraging strengths of different techniques (e.g., Hybrid LSTM-ARIMA, Ensemble Learning with Random Forest and XGBoost).

   - LSTM and ARIMA Hybrid Model: Combines ARIMA's statistical trend analysis with LSTM's ability to capture long-term dependencies for better forecasting accuracy.
   - Transformer and CNN Model: Integrates Transformers for attention-based sequence learning with CNNs for feature extraction, enhancing predictive performance.
   - Voting Ensemble (Random Forest and XGBoost with LSTM): Aggregates predictions from multiple models to improve accuracy and robustness in autoscaling decisions.

Autoscaling is a critical component of modern cloud-native applications, ensuring optimal resource utilization, cost efficiency, and high availability. By intelligently adjusting compute resources ahead of time, proactive autoscaling minimizes operational costs while maintaining seamless performance in Kubernetes-managed environments. Selecting the right autoscaling strategy depends on workload characteristics and system requirements. Reactive autoscaling is well-suited for predictable workloads where predefined thresholds effectively manage scaling. In contrast, proactive autoscaling leverages machine learning and predictive analytics to anticipate workload fluctuations, making it ideal for dynamic applications such as real-time analytics, e-commerce platforms, and IoT workloads. A hybrid approach, combining both reactive and proactive strategies, provides a balanced solution–leveraging Horizontal Pod Autoscaler (HPA) for immediate adjustments and ML-driven predictive models for long-term resource planning. By implementing an effective autoscaling strategy, organizations can enhance system resilience, reduce infrastructure costs, and ensure uninterrupted performance, making Kubernetes-based applications more adaptable to evolving workload demands.

8.3 Optimizing Allocation Using Predictive Models

In cloud computing environments, optimizing resource allocation is crucial to maintaining performance, minimizing costs, and ensuring high availability [116, 119]. Kubernetes-based custom resources integrate predictive AI scaling methods with operator reconciliation and a model control system to enhance resource management. This chapter explores how predictive models improve autoscaling efficiency, detailing their implementation, evaluation, and real-world impact.

8.3.1 Operator Reconciliation and Model Control System

A proactive autoscaling strategy relies on a structured operator reconciliation process and an intelligent model control system [111]. These components work in tandem to ensure seamless resource provisioning, workload adaptation, and system stability by integrating predictive analytics with real-time decision-making.

- Operator Reconciliation Process: The operator reconciliation mechanism plays a crucial role in maintaining system equilibrium by continuously monitoring resource metrics and making informed scaling decisions. This process involves iterative adjustments where the system evaluates its current state, predicts future workload conditions, and triggers corrective actions to prevent resource bottlenecks or over-provisioning.
  - Custom Operators for Workload Management: In cloud-native environments, custom operators serve as autonomous agents responsible for observing workload metrics such as CPU utilization, memory consumption, and network throughput. These operators analyze historical patterns and predict future demands, initiating scaling actions preemptively to maintain optimal system performance. Unlike traditional reactive approaches, this proactive mechanism minimizes latency and enhances service reliability.
- Intelligent Model Control System: The model control system acts as a decision-making framework that dynamically selects the most suitable predictive model based on workload characteristics. Given the diverse nature of workload patterns—ranging from periodic trends to bursty spikes—the ability to switch between forecasting models is essential for maintaining accuracy and efficiency.
  - Adaptive Model Selection: Various predictive models, including Transformer-based architectures, Bi-LSTM, ARIMA, and Informer-based scaling mechanisms, are continuously evaluated [117]. The system determines the most effective model in real time, considering factors such as prediction accuracy, computational efficiency, and adaptability to workload fluctuations. By dynamically adjusting the scaling strategy, the model control system ensures

optimal resource utilization while mitigating the risks associated with under-provisioning or excessive resource allocation.

- Asynchronous Execution with Coroutines: To handle large-scale workloads efficiently, asynchronous execution techniques are employed, allowing concurrent task processing without blocking operations. This approach significantly enhances the responsiveness and scalability of the system.
  - Coroutines for Efficient Resource Management: Coroutines enable lightweight, non-blocking execution of scaling tasks, ensuring that predictive autoscaling operations do not introduce unnecessary delays. By leveraging asynchronous task execution, the system can manage multiple scaling requests in parallel, enhancing responsiveness in dynamic cloud environments.

By integrating these components, the proactive autoscaling strategy ensures that workload fluctuations are addressed with minimal latency, maintaining a balance between resource efficiency and application performance.

8.4 Performance Evaluation

Evaluating the performance of predictive autoscaling involves assessing how effectively the system anticipates workload fluctuations and dynamically adjusts resources to maintain optimal service levels. This assessment considers key performance indicators, including error metrics and resource utilization, ensuring that predictive autoscaling strategies are both accurate and efficient.

- Workload Analysis: To analyze the behavior of predictive autoscaling, historical workload traces from large-scale web applications are examined. These traces provide insights into real-world fluctuations in request patterns, offering a diverse range of traffic intensities that serve as a basis for testing the adaptability of predictive scaling mechanisms. By understanding workload variations, the system can anticipate demand surges and declines, allowing for timely adjustments in resource provisioning. Workload analysis also involves categorizing traffic patterns, and distinguishing between periodic trends, bursty behaviors, and unexpected spikes. This classification aids in fine-tuning predictive models, ensuring that they can generalize effectively across different usage scenarios. By leveraging machine learning techniques, the system can recognize recurring trends and anomalies, leading to more informed scaling decisions.
- Accuracy of Predictive Scaling: One of the critical aspects of evaluating predictive autoscaling is measuring the accuracy of workload forecasting. Traditional reactive scaling mechanisms adjust resources only after demand fluctuations occur, often resulting in delays and inefficient resource usage. In contrast, predictive scaling relies on forecasting techniques to proactively allocate resources before demand peaks or drops. To quantify forecasting accuracy, various loss functions and error metrics are used. These metrics assess how closely the

predicted workload aligns with actual usage patterns. Lower error values indicate a more precise forecasting model, leading to improved efficiency in autoscaling decisions. By reducing prediction errors, the system minimizes unnecessary resource provisioning while ensuring sufficient capacity to handle demand surges.

- Evaluation of Resource Utilization Apart from forecasting accuracy, the efficiency of predictive autoscaling is evaluated based on resource utilization metrics. These metrics provide insights into how well the system balances resource allocation with actual demand, preventing both over-provisioning and under-provisioning scenarios.
  - Under-Provisioning Impact: When resource allocation falls short of demand, applications may experience performance degradation, increased latency, and potential service disruptions. Measuring the frequency and duration of under-provisioning instances helps assess the effectiveness of predictive scaling in mitigating such risks.
  - Elasticity and Scaling Responsiveness: The ability of the system to dynamically adjust resources in response to workload changes is a key performance indicator. Faster and more accurate scaling improves overall system responsiveness, ensuring that applications maintain high availability and performance levels.
  - Adaptation to Workload Variability: Predictive scaling mechanisms should demonstrate adaptability across different workload conditions, including steady-state operations, peak traffic periods, and unexpected demand surges. Evaluating system behavior under diverse scenarios helps validate its robustness and reliability.

By systematically analyzing these performance aspects, predictive autoscaling can be refined to achieve a balance between efficiency and cost-effectiveness, enhancing overall cloud resource management.

8.5 Real-World Implementation and Performance Evaluation

To validate proactive autoscaling in production, its integration into Kubernetes and Docker environments is explored, along with a real-world case study.

8.5.1 Kubernetes and Docker

Integration Kubernetes and Docker Integration enables seamless containerized application deployment and management by leveraging predictive autoscaling techniques. By integrating proactive scaling strategies, Kubernetes orchestrates container workloads efficiently, while Docker ensures lightweight, scalable execution, optimizing resource utilization in dynamic cloud environments.

- Kubernetes Orchestration: Kubernetes, as a container orchestration platform, plays a crucial role in managing and automating the deployment, scaling, and operation of containerized applications. In the context of proactive autoscaling, predictive models are integrated with Kubernetes to enhance its decision-making process regarding pod scaling. These models analyze historical and real-time workload data to forecast future demand, ensuring that the system provisions the necessary resources in advance. By leveraging machine learning algorithms or time-series forecasting techniques, Kubernetes can dynamically adjust the number of running pods to match anticipated demand, mitigating the risks associated with sudden traffic surges or unexpected workload variations.
- Docker Container Scaling: Docker containers form the foundation of microservices-based architectures, where each container encapsulates an application or service with its dependencies. Proactive autoscaling in a Docker environment is driven by workload predictions that enable real-time scaling of containers. Through integration with Kubernetes' Horizontal Pod Autoscaler (HPA) or Vertical Pod Autoscaler (VPA), predictive models can inform scaling decisions, ensuring that virtual machines (VMs) and underlying infrastructure efficiently adapt to changing demands. This approach optimizes resource utilization, preventing over-provisioning and minimizing operational costs. Furthermore, container-level autoscaling enhances application resilience, maintaining performance stability even during unpredictable traffic fluctuations.

8.5.2 Case Study: Dynamic Allocation in Cloud Computing

To demonstrate the effectiveness of proactive autoscaling, a case study was conducted using the Monitor-Analyze-Plan-Execute (MAPE) framework [114]. This framework facilitates real-time decision-making by continuously monitoring workloads, analyzing trends, planning resource allocations, and executing scaling actions in an automated manner.

- Implementation of the MAPE Framework: To further enhance proactive autoscaling, the MAPE (Monitor-Analyze-Plan-Execute) framework was implemented within a cloud computing environment [120]. The MAPE (Monitor, Analyze, Plan, Execute) framework is a widely used approach in autonomous computing systems for adaptive resource management. It consists of four key phases:
  - The Monitor component Continuously gathers system metrics such as CPU and memory usage, network throughput, and request rates to assess system performance.

   - Analyze—Processes collected data using predictive analytics and machine learning to identify trends, anomalies, and potential workload fluctuations.
   - Plan—Formulates an optimal resource management strategy based on the analyzed data to ensure efficient scaling and workload distribution.
   - Execute—Implements the planned actions by dynamically adjusting resource provisioning and deallocation to maintain system performance and minimize operational costs.

  This cyclic process enables automated and adaptive decision-making in cloud and edge computing environments. This structured approach ensures that cloud applications remain responsive and cost-effective, even under unpredictable workload variations. The next chapter will delve deeper into advanced predictive modeling techniques that further refine autoscaling decisions and enhance system performance.
- Evaluation Under Dynamic Traffic Conditions: Performance tests were conducted under varying traffic loads to assess the efficiency of proactive autoscaling. The evaluation focused on key performance indicators (KPIs) such as response time, resource utilization, and cost efficiency. The results highlighted significant improvements in system performance compared to traditional reactive autoscaling approaches.

8.6 Key Finding

Autoscaling is a crucial technique in cloud computing that dynamically adjusts computational resources based on workload demands. Traditional reactive autoscaling responds to changes after they occur, often leading to service delays and inefficient resource utilization. In contrast, proactive autoscaling leverages predictive models to anticipate workload fluctuations and allocate resources in advance. This proactive approach significantly enhances system performance, reduces operational costs, and improves user experience. Below is a detailed breakdown of the key findings regarding predictive autoscaling.

8.6.1 Reduced Latency

Latency refers to the delay experienced by users when accessing an application or service. In traditional cloud environments, when a surge in demand occurs, reactive scaling methods take time to recognize the increase and provision additional resources. This delay can result in service degradation, slow response times, and potential downtime.

- How Predictive Scaling Reduces Latency: Anticipation of Demand Surges: Predictive autoscaling uses historical data and machine learning algorithms

to forecast spikes in demand before they happen. This allows the system to provision additional resources in advance, ensuring that the required capacity is available when needed.
- Minimization of Service Interruptions: Since computational resources are already allocated before a workload surge, there is no waiting time for new instances to start, reducing application response delays. Improved Performance Stability: By maintaining a consistent resource supply, predictive scaling prevents bottlenecks that typically occur in reactive autoscaling, ensuring seamless user interactions and improved service reliability.

For example, an e-commerce platform expecting high traffic during a sale event can use predictive autoscaling to pre-allocate additional server capacity, ensuring smooth operations without lag or downtime.

8.6.2 Optimized Cloud Expenses

Cloud infrastructure costs are a significant concern for organizations, especially when dealing with fluctuating workloads. Many businesses struggle with over-provisioning (allocating more resources than necessary) and under-provisioning (insufficient resources, leading to performance issues).

- How Predictive Scaling Optimizes Cloud Costs: Efficient Resource Allocation: Instead of maintaining a constant pool of resources based on worst-case scenarios, predictive autoscaling dynamically adjusts resource provisioning based on actual demand predictions. This prevents unnecessary resource allocation, leading to cost savings.
- Reduction in Over-Provisioning: Many organizations provision more resources than needed to avoid performance degradation. Predictive autoscaling ensures just-in-time resource allocation, avoiding excess costs while still maintaining application performance.
- Elimination of Waste: By only using resources when they are needed, cloud expenses are reduced without negatively impacting performance. This is particularly beneficial for businesses with variable workloads, such as streaming services or financial trading platforms, where demand can fluctuate unpredictably.

For instance, a video streaming platform might see higher traffic during prime hours. Predictive scaling ensures that the necessary computational power is allocated before peak times and scaled down during off-peak hours, reducing unnecessary cloud expenses.

8.6.3 *Improved Response Times*

Response time is a critical factor in user experience. Applications that suffer from slow response times due to delayed scaling can frustrate users, leading to lower customer satisfaction and potential revenue loss.

- How Predictive Scaling Improves Response Times: Preemptive Scaling for Smoother Operations: Since resources are allocated in advance, applications can handle traffic spikes more efficiently, preventing sudden slowdowns.
- Reduction of System Bottlenecks: Traffic spikes often cause system congestion, leading to processing delays. Predictive autoscaling prevents these bottlenecks by ensuring that the necessary computational resources are available beforehand.
- Consistent and Stable Performance: Unlike reactive methods, where the system takes time to recognize a demand spike and then allocate resources, predictive scaling ensures continuous system stability even during unpredictable workload variations.

For example, an online banking system that experiences sudden transaction surges during salary deposit times benefits from predictive scaling. The system forecasts high-demand periods and scales resources accordingly, ensuring fast transaction processing without delays.

8.6.4 *Practical Implementation in Cloud Environments*

Predictive autoscaling is highly effective when integrated into cloud environments such as Kubernetes and Docker, which manage containerized applications. The MAPE (Monitor, Analyze, Plan, Execute) framework is commonly used to implement predictive scaling by continuously analyzing system metrics and adjusting resources dynamically.

- Key Benefits of this Integration: Seamless Cloud Resource Management: Kubernetes and Docker allow efficient scaling of containerized applications. Predictive scaling enhances this by ensuring optimal container deployment and resource utilization.
- Enhanced Resilience: By preparing for workload surges in advance, systems can maintain high availability and performance without disruption.
- Intelligent and Adaptive Cloud Infrastructure: Moving beyond reactive scaling methods enables organizations to create self-optimizing cloud environments that adjust resources dynamically based on real-time and predicted data.

For instance, a SaaS-based application running on Kubernetes can leverage predictive autoscaling to maintain optimal performance across different time zones, dynamically adjusting resources based on user activity.

Overall, the adoption of predictive autoscaling in cloud computing presents significant advantages over traditional reactive methods. By reducing latency, optimizing cloud expenses, and improving response times, organizations can ensure better performance, cost efficiency, and user experience. The integration of predictive models with Kubernetes, Docker, and the MAPE framework further enhances cloud infrastructure management, making it more intelligent, adaptive, and resilient. This shift towards proactive autoscaling represents a critical step in the evolution of cloud resource allocation, enabling more efficient and cost-effective cloud environments for modern applications.

8.7 Advancing Predictive Autoscaling for Cloud and Edge Computing

Predictive autoscaling plays a crucial role in optimizing cloud resource management by addressing the limitations of traditional reactive methods. Reactive autoscaling, which relies on predefined thresholds, often results in delayed responses to workload fluctuations. This delay can cause system inefficiencies, increased latency, and higher operational costs. In contrast, predictive autoscaling leverages advanced forecasting techniques to anticipate workload demands, ensuring that resources are allocated proactively. By preemptively adjusting computing power based on expected usage patterns, predictive scaling significantly enhances system responsiveness, minimizes performance degradation, and prevents resource wastage. One of the primary benefits of predictive autoscaling is its ability to reduce latency. By provisioning resources before a surge in demand occurs, applications maintain optimal performance levels, avoiding service disruptions. Additionally, this approach optimizes cloud expenses by preventing over-provisioning and ensuring efficient utilization of computing power. Organizations can maintain cost-effectiveness without sacrificing service quality. Furthermore, predictive autoscaling enhances overall system stability, enabling applications to handle traffic spikes more smoothly, reducing the likelihood of bottlenecks, and improving user experience. Future advancements in predictive autoscaling should focus on refining forecasting models using reinforcement learning. By continuously learning from past scaling decisions, reinforcement learning can improve the accuracy of predictions and optimize resource provisioning dynamically. Another promising direction is the development of hybrid autoscaling strategies that combine proactive and reactive approaches. A hybrid model would allow systems to benefit from both real-time adjustments and long-term forecasting, ensuring a more adaptive and efficient scaling mechanism. Additionally, expanding predictive autoscaling to multi-cloud and edge computing environments will enhance resource distribution across diverse infrastructures. As cloud and edge systems grow more complex, optimizing resource allocation in distributed environments will become critical for maintaining performance and reliability. By integrating advanced machine learning techniques and hybrid scaling strategies, predictive autoscaling can evolve into a more intelligent and adaptable solution, paving the way for next-generation cloud and edge computing management.

8.8 Key Takeaways

- Autoscaling in Kubernetes optimizes resource allocation and ensures efficient workload management by dynamically adjusting resources.
- Reactive autoscaling scales resources based on real-time metrics such as CPU and memory usage, responding to current system load.
- Proactive autoscaling anticipates future demand using predictive models, enabling preemptive scaling to enhance system stability and performance.
- Predictive autoscaling improves resource efficiency by reducing latency, optimizing cloud expenses, and enhancing response times.
- Operator reconciliation and model control systems refine autoscaling decisions by continuously adjusting to changing workloads.
- Case studies on Kubernetes and Docker demonstrate the practical benefits of autoscaling strategies in real-world scenarios.
- The future of autoscaling in cloud and edge computing requires AI-driven adaptive strategies to handle dynamic workload demands effectively.

8.9 Summary

This chapter examines autoscaling mechanisms in Kubernetes, comparing reactive and proactive approaches to resource management. Reactive autoscaling, which adjusts resources based on real-time metrics like CPU and memory usage, often leads to inefficiencies due to delayed responses and over-provisioning. In contrast, proactive autoscaling leverages predictive models to anticipate workload fluctuations, enabling resource allocation before demand surges and improving system stability. By integrating predictive models with operator reconciliation and model control systems, Kubernetes can optimize resource provisioning, reducing latency, minimizing cloud expenses, and enhancing response times. A real-world case study demonstrates the practical benefits of proactive scaling in Kubernetes and Docker environments, showcasing its effectiveness in managing dynamic cloud workloads. Performance evaluations highlight that predictive scaling significantly enhances workload handling by preventing service disruptions and ensuring optimal resource utilization. Key insights from this chapter emphasized the benefits of predictive autoscaling, including reduced latency, optimized cloud expenses, and improved response times. The discussion concluded with an outlook on advancing predictive autoscaling for cloud and edge computing, stressing the importance of AI-driven models for handling heterogeneous and fluctuating workloads. Future advancements in autoscaling focus on integrating reinforcement learning to refine predictive models and developing hybrid strategies that combine proactive and reactive methods for greater adaptability. Expanding autoscaling to multi-cloud and edge computing environments will further optimize distributed resource allocation, ensuring efficiency across diverse infrastructures. Ultimately, this chapter underscores the need for intelligent predictive autoscaling to improve performance, reduce costs, and enhance the resilience of cloud computing systems.

8.10 Review Questions

- What are the fundamental differences between reactive and proactive autoscaling in Kubernetes?
- How does Kubernetes' built-in Horizontal Pod Autoscaler (HPA) facilitate reactive scaling?
- What are the main categories of autoscaling strategies, and how do they differ in approach?
- How does reactive autoscaling respond to real-time workload changes, and what are its limitations?
- What are the key advantages of proactive autoscaling over reactive methods?
- How do predictive models enhance resource allocation in proactive autoscaling?
- Explain the role of the Operator Reconciliation Loop in Kubernetes autoscaling.
- How does a Model Control System improve decision-making in proactive autoscaling?
- What metrics are crucial for performance evaluation when comparing reactive and proactive approaches?
- How do Kubernetes and Docker work together to enable efficient autoscaling in cloud environments?
- What were the key findings from the case study on dynamic allocation in cloud computing?
- How did predictive autoscaling contribute to reduced latency and optimized cloud expenses?
- What practical steps are needed to implement predictive autoscaling in cloud environments?
- How can edge computing benefit from proactive autoscaling strategies?
- What are the future research directions for advancing predictive autoscaling in Kubernetes?

Chapter 9
Intelligent Autoscaling with the MAPE Framework

Abstract This chapter introduces intelligent autoscaling in Kubernetes using the MAPE (Monitor-Analyze-Plan-Execute) framework and its extended form, MAPE-K, which incorporates knowledge-driven decision-making. It begins with an explanation of MAPE components and their application in reactive autoscaling, highlighting both the advantages of responsiveness and the limitations of delayed reaction times. The proactive approach within the MAPE framework is then discussed, demonstrating how predictive analysis and planning can anticipate workload changes for optimized scaling. The MAPE-K extension is examined in terms of its uses, benefits, and challenges, particularly its role in adaptive, knowledge-based resource management. Performance evaluation is supported by key metrics, along with real-world implementations and case studies that show practical applications of the framework. The chapter concludes with a discussion of advanced research directions in intelligent autoscaling, emphasizing how MAPE and MAPE-K approaches can enhance cloud-native and edge computing environments by unifying automation, prediction, and adaptive orchestration.

Keywords MAPE framework · MAPE-K · Intelligent autoscaling · Reactive scaling · Proactive scaling · Monitor-analyze-plan-execute · Cloud-native orchestration · Automation · Adaptive resource management

Intelligent autoscaling is a critical component of cloud computing that ensures efficient resource allocation based on workload demands [110]. The MAPE framework (Monitor, Analyze, Plan, and Execute) provides a structured approach to implementing intelligent autoscaling, enabling adaptive and self-optimizing systems (see Fig. 9.1). This framework, originally designed for autonomic computing, serves as a self-adaptive system model that automates decision-making for dynamic resource management [121]. In intelligent autoscaling, MAPE continuously monitors workload fluctuations, analyzes usage patterns, plans appropriate scaling actions, and executes them in real-time [122]. By integrating machine learning (ML) and predictive analytics, MAPE-based autoscaling enhances resource utilization,

B. Kumar et al., *Modern Kubernetes: From Core Concepts to Intelligent Autoscaling for Cloud Applications*, Studies in Autonomic, Data-driven and Industrial Computing, https://doi.org/10.1007/978-3-032-12972-7_9

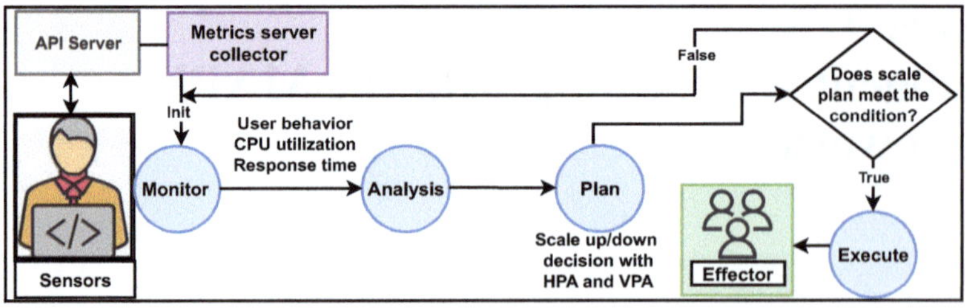

Fig. 9.1 Architecture of MAPE framework

cost-efficiency, and system reliability. One of the key advantages of MAPE-based autoscaling is its ability to proactively adjust resources, preventing performance degradation while avoiding unnecessary scaling. This results in reduced operational costs, improved system efficiency, and higher availability of cloud services. Additionally, automation minimizes human intervention, making resource management more scalable and efficient. However, there are also challenges associated with implementing MAPE in intelligent autoscaling. Accurate workload prediction remains a significant challenge, as unforeseen traffic spikes or underutilization can lead to inefficient scaling decisions [123]. Computational overhead from continuous monitoring and analysis may impact system performance, especially in high-load environments. Moreover, integrating heterogeneous cloud resources across multi-cloud or hybrid environments complicates decision-making and execution. Despite these challenges, the benefits of MAPE-based autoscaling make it a crucial strategy for cloud-native applications. It ensures adaptive resource provisioning, enhances system responsiveness, and supports sustainable cloud operations by optimizing energy consumption. As cloud computing continues to evolve, advancements in AI-driven analytics and edge computing will further enhance the effectiveness of MAPE-based intelligent autoscaling.

9.1 MAPE Component

The MAPE Framework (Monitor, Analyze, Plan, Execute) is a self-adaptive model that enables intelligent autoscaling in cloud computing [124]. It consists of four key phases that work together to automate scaling decisions. In the diagram illustrating the working of the MAPE framework (see Fig. 9.1), this phase collectively connects to the master node, which was already shown in the previous chapter. Additionally, there is a connection to the API server that collects metrics through the default Kubernetes approach or software tools like Prometheus for faster data collection. After collecting the metrics, they are processed through the monitoring system, utilizing the development side and referencing the sensor side as input for validation and testing. The data is then forwarded to the analysis, planning, and execution phases, ultimately impacting the user or real-life platform. Each phase is used separately, as described below.

- **Monitor (M):** The Monitor component plays a crucial role in the MAPE (Monitor, Analyze, Plan, Execute) framework, continuously observing the system and collecting real-time performance metrics. These metrics include CPU usage, memory consumption, disk I/O, network traffic, application response times, and request loads. By gathering this data, the system can detect workload fluctuations and make informed scaling decisions. To achieve effective monitoring, tools such as Prometheus, Grafana, and CloudWatch are widely used in cloud environments. These tools help capture, visualize, and analyze system metrics, ensuring that resources are utilized efficiently. However, a major challenge in this phase is maintaining low-latency data collection while minimizing the computational overhead that could impact overall system performance. By continuously monitoring resource utilization and application behavior, this component helps identify trends, detect anomalies, and provide the necessary input for the Analyze phase. Accurate and efficient monitoring is essential for ensuring that the system remains responsive to changing workload demands, enabling proactive resource management and intelligent autoscaling.
- **Analyze (A):** The Analyze phase is responsible for evaluating the collected monitoring data to determine whether scaling actions are necessary. This phase helps identify patterns, trends, and potential performance issues that could impact system stability and efficiency. By leveraging machine learning models, statistical analysis, and rule-based heuristics, the system can detect anomalies, predict future demand, and recognize bottlenecks before they cause performance degradation. Several techniques are used in this phase to analyze data effectively. Threshold-based analysis involves setting predefined limits, such as triggering scaling when CPU usage exceeds 80%. More advanced approaches include machine learning models for predictive analysis and anomaly detection, which allow the system to anticipate resource demands in advance. Additionally, statistical methods such as moving averages and time-series forecasting help in trend analysis and workload prediction. Despite its advantages, the Analyze phase presents challenges, particularly in achieving accurate predictions. Workload variations can be highly dynamic and unpredictable, making it difficult to determine the ideal scaling action. Incorrect predictions can lead to inefficient resource allocation, either causing performance bottlenecks or unnecessary scaling operations that increase operational costs. Overcoming these challenges requires robust forecasting models and continuous refinement of predictive algorithms to improve decision-making accuracy in intelligent autoscaling.
- **Plan (P):** The Plan phase is responsible for formulating scaling decisions based on the insights derived from the Analyze phase. This phase determines the most appropriate scaling strategy to ensure optimal resource utilization, system performance, and cost efficiency [125]. Scaling decisions can be reactive, where resources are adjusted in response to real-time fluctuations, or predictive, where future demand is anticipated using forecasting techniques such as ARIMA models or deep learning. A hybrid approach that integrates both reactive and predictive scaling can improve efficiency by reducing unnecessary scaling

actions and preventing performance bottlenecks. In this phase, the system decides whether to:

- Scale-out—Add more instances or pods to handle the increased workload.
- Scale in—Remove unused resources to optimize costs.
- Scale up/down—Adjust resource limits per instance to match workload demands.
- Load balance—Redistribute workloads efficiently to prevent resource congestion.

One of the biggest challenges in the Plan phase is achieving the right balance between performance, cost, and energy efficiency. Frequent scaling actions can cause oscillations, where the system continuously scales up and down, leading to instability and unnecessary resource consumption. To mitigate this, intelligent autoscaling systems must incorporate optimization mechanisms that ensure stable and effective scaling strategies while minimizing disruptions to ongoing workloads.

- **Execute (E):** The Execute phase is responsible for implementing the scaling decisions made during the Plan phase. This involves dynamically adjusting resources to meet changing workload demands while maintaining system stability and performance. Scaling actions in this phase can include scaling out by adding new instances, scaling in by removing underutilized instances, scaling up or down by modifying the resource limits of existing instances, or redistributing traffic through load balancing to ensure an even distribution of workloads. Automation plays a crucial role in the Execute phase, with orchestration tools such as Kubernetes Horizontal Pod Autoscaler (HPA), Vertical Pod Autoscaler (VPA), AWS Auto Scaling, and OpenStack Heat managing resource adjustments seamlessly. These tools continuously monitor system performance and execute scaling actions without requiring manual intervention, ensuring efficiency and responsiveness. A key challenge in this phase is ensuring that scaling actions are carried out without disrupting ongoing services. Poorly executed scaling operations can lead to temporary downtime, degraded performance, or inefficient resource allocation. To overcome these challenges, intelligent autoscaling systems must incorporate mechanisms that allow for smooth transitions, such as graceful scaling techniques and proactive load redistribution, ensuring that performance remains stable even as resources are adjusted dynamically.

By integrating the MAPE (Monitor, Analyze, Plan, Execute) framework, intelligent autoscaling enhances resource utilization, reduces operational costs, and improves application performance. This structured approach ensures that cloud-native applications remain highly available, responsive, and efficient, even in fluctuating workload conditions. By continuously monitoring system metrics [126], analyzing performance trends, planning scaling strategies, and executing them automatically, the MAPE framework minimizes the need for manual intervention, making resource management more autonomous and adaptive. One of the key advantages of MAPE-based autoscaling is its ability to balance performance and cost-effectiveness. By leveraging real-time monitoring and predictive analytics, the system can anticipate resource demands, reducing the risk of both under-provisioning (which leads to performance degradation) and over-provisioning (which increases unnecessary costs). Additionally, automation ensures that resources are allocated dynamically, responding to workload changes without requiring constant human oversight. Despite its benefits, implementing the MAPE framework for autoscaling presents several challenges. Accurate data collection and analysis are critical, as errors in monitoring or prediction can lead to inefficient scaling decisions. Latency in scaling actions is another challenge, as delays in resource allocation may temporarily degrade application performance. Furthermore, balancing energy efficiency and performance remains an ongoing concern, particularly in large-scale cloud environments where frequent scaling actions can introduce system instability.

Overall, the MAPE framework provides a structured and intelligent approach to cloud autoscaling, ensuring that applications operate efficiently while maintaining reliability and cost efficiency. As cloud technologies evolve, integrating machine learning and AI-driven decision-making within the MAPE cycle can further enhance its capabilities, making cloud resource management even more adaptive and proactive.

9.2 MAPE Framework for Reactive Autoscaling

MAPE Framework for Reactive Autoscaling Reactive autoscaling is a widely used technique in cloud environments, where scaling decisions are made based on real-time system metrics [108]. This approach responds to changes in resource demand only after they occur, making it effective for handling predictable and gradual workload fluctuations. The MAPE (Monitor, Analyze, Plan, Execute) framework plays a crucial role in implementing reactive autoscaling by automating the decision-making process based on predefined rules.

- Monitor (Real-Time Data Collection): The monitoring phase in reactive autoscaling continuously tracks key performance indicators (KPIs) such as CPU utilization, memory consumption, disk I/O, and network traffic. This data is collected using monitoring tools like Prometheus, Grafana, AWS CloudWatch, and Kubernetes Metrics Server. The goal is to detect deviations from expected workload patterns and provide real-time insights into system performance.
  - Key Metrics Monitored: CPU usage (%): Determines when to scale based on threshold-based triggers (e.g., scaling out if CPU > 80%). Memory consumption (MB/GB): Ensures sufficient memory availability for running workloads. Network throughput (MBps): Helps identify network congestion that may require load balancing. **Response time (ms):** Measures application performance and user experience.
  - Challenges in Monitoring: Latency in data collection: Monitoring tools must provide near real-time metrics to prevent delays in scaling actions. Overhead

from excessive monitoring: Continuous data collection can introduce additional resource consumption, affecting system performance.

- Analyze (Identifying Scaling Needs): Once monitoring data is collected, the Analyze phase processes this information to determine whether scaling actions are necessary. Reactive autoscaling typically relies on threshold-based rules, meaning that when a certain metric crosses a predefined limit, a scaling action is triggered.
  - Common Analytical Approaches: Threshold-based rules: Example: If CPU utilization > 80%, add a new instance; if CPU < 30%, remove an instance.
  - Statistical analysis: Moving averages or historical data comparisons help refine scaling thresholds.
  - Anomaly detection: Detects unusual spikes that may indicate sudden demand surges.

  One of the primary challenges in the Analyze phase of reactive autoscaling is defining optimal scaling thresholds. Fixed threshold values, such as scaling out when CPU utilization exceeds 80% or scaling in when it drops below 30%, may not be suitable for all workloads. Cloud environments are highly dynamic, with resource demands fluctuating based on user activity, application behavior, and external factors. A static threshold-based approach may either overprovision resources, leading to increased operational costs, or under-provision them, resulting in performance degradation. Adaptive thresholding techniques, such as moving averages or dynamic baselines, can help mitigate this issue but require additional computational overhead and fine-tuning. Another challenge is the occurrence of false positives, where short-lived spikes in resource utilization trigger unnecessary scaling actions. In many cases, workload surges are transient and do not require additional resources. However, if the autoscaler reacts too quickly, it may introduce unnecessary scaling operations, leading to resource oscillation—a situation where instances are frequently added and removed, causing instability in the system. To prevent this, hysteresis mechanisms and cool-down periods are often implemented to delay scaling decisions until a sustained trend is detected. However, finding the right balance between responsiveness and stability remains a significant challenge in the analysis phase of reactive autoscaling.
- Plan (Determining Scaling Actions): In this phase, the system decides what type of scaling action should be taken based on the analysis results. The goal is to balance performance, cost-efficiency, and stability while ensuring that resources meet workload demands. The Plan phase in the MAPE framework involves determining the appropriate scaling strategy based on workload analysis while addressing key challenges in decision-making. One major challenge is avoiding rapid oscillations, where frequent scaling actions destabilize the system, causing unnecessary resource allocation and performance fluctuations. This issue can be mitigated by implementing hysteresis mechanisms and cool-down periods to ensure that scaling decisions are triggered only when sustained trends are

detected. Another critical challenge is balancing cost and performance, as over-scaling leads to increased operational expenses, while under-scaling degrades application responsiveness. To optimize this trade-off, predictive scaling techniques and adaptive thresholding can be employed, enabling smarter scaling decisions that align with workload variations while maintaining cost efficiency.

- Execute (Implementing Scaling Actions): Once a scaling decision is made, the Execute phase applies the necessary resource adjustments. This process is typically automated using orchestration tools like Kubernetes Horizontal Pod Autoscaler (HPA), AWS Auto Scaling, and OpenStack Heat.
  - The Execute phase of the MAPE framework is responsible for implementing scaling actions determined during the planning stage. Various execution mechanisms are used to dynamically adjust resources based on demand. Kubernetes Horizontal Pod Autoscaler (HPA) and Vertical Pod Autoscaler (VPA) automatically modify the number of running pods or adjust container resource limits to optimize performance. Similarly, AWS Auto Scaling Groups enable automatic provisioning and termination of EC2 instances, ensuring that cloud resources scale dynamically with workload fluctuations. Additionally, cloud load balancers play a crucial role in distributing incoming traffic across multiple instances, preventing resource congestion and maintaining system stability. Despite its advantages, the execution phase presents several challenges.
  - Challenges: One key issue is ensuring seamless scaling, where scaling actions must be executed without causing downtime or degrading application performance. This requires careful orchestration to avoid resource contention and maintain system responsiveness during scaling transitions. Another challenge is avoiding unnecessary scaling, as short-lived spikes in workload demand should not trigger excessive resource allocation, leading to inefficiencies and increased costs. Implementing cool-down periods and intelligent scaling thresholds can help mitigate this issue, ensuring that scaling actions are based on sustained trends rather than transient fluctuations.

9.2.1 Advantages of Reactive Autoscaling

Despite its limitations, reactive autoscaling remains a widely used approach in cloud resource management due to its simplicity, reliability, and ease of integration. One of its primary advantages is its ease of implementation, as threshold-based rules are straightforward to configure and can be seamlessly integrated with existing cloud platforms like Kubernetes, AWS, and Azure. This simplicity makes it an attractive choice for organizations looking to automate resource allocation without requiring complex predictive models. Additionally, reactive autoscaling is highly effective for predictable workloads, particularly for applications with steady traffic patterns that do not require sophisticated scaling strategies. By automatically adjusting resources in response to demand fluctuations, it ensures automated resource management, reducing the need for manual intervention and enhancing operational efficiency. This dynamic resource allocation helps optimize cloud infrastructure costs while maintaining application performance.

9.2.2 Challenges and Limitations of Reactive Autoscaling

While reactive autoscaling provides a fundamental level of automation and resource optimization, it has several inherent drawbacks that can impact performance and efficiency. One of the key challenges is its slow response to sudden workload spikes, as scaling decisions are only made after resource utilization surpasses a predefined threshold. This reactive nature can lead to temporary performance degradation, as new resources may take time to spin up, causing delays in handling increased traffic. Additionally, there is a risk of under-provisioning, where resource allocation occurs too slowly, leading to service delays or application failures during high-demand periods. Another challenge is oscillations in scaling, where frequent scaling actions—such as repeatedly adding and removing instances—can cause system instability and increase operational costs. Finally, limited adaptability to dynamic workloads poses a significant constraint, as predefined scaling rules may not effectively handle unpredictable traffic patterns. To address these limitations, advanced autoscaling strategies, such as predictive and hybrid autoscaling, can be explored to enhance responsiveness and efficiency.

In conclusion, Reactive autoscaling, driven by the MAPE framework, is a fundamental mechanism for cloud environments that require automated, rule-based resource management. By continuously monitoring system metrics, analyzing workload patterns, planning scaling actions, and executing resource adjustments, it ensures operational efficiency and system stability. This approach effectively optimizes resource utilization while minimizing manual intervention, making it a practical solution for applications with predictable traffic patterns. However, its delayed response to sudden workload fluctuations remains a significant limitation, making it less suitable for applications with highly dynamic or unpredictable resource demands. To overcome these challenges, future enhancements incorporating predictive analytics, AI-driven decision-making, and hybrid scaling approaches can improve its efficiency and adaptability. By integrating these advanced techniques, cloud resource management can become more intelligent, proactive, and resilient, ensuring optimal performance and cost-effectiveness in increasingly complex cloud environments.

9.3 MAPE Framework for Proactive Autoscaling

MAPE Framework for Proactive Autoscaling Proactive autoscaling anticipates future workload demands and adjusts resources in advance using predictive techniques such as time-series forecasting, reinforcement learning, and hybrid machine learning models [118]. Unlike reactive autoscaling, which responds to changes after they occur, proactive autoscaling ensures optimal resource allocation before demand spikes or drops, enhancing system stability and cost efficiency. The MAPE (Monitor, Analyze, Plan, Execute) framework provides a structured methodology for implementing proactive autoscaling in cloud environments [108].

- Monitor: The Monitor phase continuously collects performance metrics, including CPU utilization, memory consumption, and network traffic. Unlike reactive monitoring, which focuses on real-time resource usage, proactive monitoring stores and analyzes historical data to identify long-term trends. This enables more accurate forecasting, but it also introduces challenges such as increased data storage requirements and computational overhead for real-time processing.
- Analyze: In the Analyze phase, predictive models process collected data to forecast future workload fluctuations. Machine learning algorithms such as ARIMA, LSTMs, and reinforcement learning-based models analyze patterns and trends to anticipate resource demands. This phase reduces scaling latency and improves cost efficiency, but it relies heavily on the accuracy of the prediction models. Errors in forecasting can lead to over-provisioning, increasing costs, or under-provisioning, causing performance degradation.
- Plan: The Plan phase formulates scaling strategies based on predictive insights. The system determines whether to scale out by adding resources before a predicted spike, scale in to reduce excess capacity during low demand, or adjust the resource limits of existing instances. Proactive planning enhances performance stability, but it also increases complexity as decisions must balance prediction accuracy with real-time operational needs, minimizing unnecessary scaling actions.
- Execute: In the Execute phase, the planned scaling actions are implemented using orchestration tools like Kubernetes HPA/VPA, AWS Auto Scaling, or OpenStack Heat. These tools automate resource adjustments, ensuring seamless scaling before workload fluctuations impact performance. The main challenge in execution is avoiding unnecessary scaling due to short-lived demand variations while ensuring that scaling actions do not disrupt ongoing services.

Proactive autoscaling offers several advantages, including enhanced performance stability, cost efficiency, and reduced manual intervention. However, challenges such as forecasting inaccuracies, computational overhead, and implementation complexity must be addressed for effective deployment. The integration of AI-driven scaling and reinforcement learning techniques can further improve its adaptability, making cloud resource management more intelligent and autonomous.

9.4 Monitor-Analyze-Plan-Execute-Knowledge (MAPE-K)

The Monitor-Analyze-Plan-Execute-Knowledge (MAPE-K) framework is integral to developing self-adaptive systems across various domains, including cloud infrastructure management, edge computing, and Kubernetes-based microservices [127]. Its structured approach facilitates adaptive resource scaling, enhances performance, and reduces operational costs. However, implementing the MAPE-K framework presents challenges related to model accuracy, balancing resource provisioning, and managing real-time execution complexities.

9.4.1 Uses

The MAPE-K framework is widely applied in cloud computing, edge environments, and microservices orchestration, enabling automated and intelligent resource management. Its adaptability makes it suitable for dynamic and complex computing infrastructures.

- Cloud Infrastructure Management: The MAPE-K framework enables autonomic resource management in cloud environments by continuously monitoring system performance, analyzing data to detect anomalies, planning appropriate responses, and executing actions to maintain optimal operation. This self-adaptive approach ensures that cloud services meet Quality-of-Service (QoS) requirements and adhere to Service-Level Agreements (SLAs).
- Edge Computing: In edge computing, integrating adaptable microservices within a MAPE-K loop allows for real-time adjustments to service variants, balancing QoS metrics such as execution time and computational quality. This integration is crucial for managing the dynamic and resource-constrained nature of edge environments.
- Kubernetes-based Microservices: Kubernetes facilitates the orchestration of containerized applications, and incorporating the MAPE-K framework enhances its capability to manage microservices architectures effectively. This integration supports automated scaling, load balancing, and self-healing, thereby improving the resilience and efficiency of microservices deployments.

9.4.2 Advantages

By leveraging real-time monitoring, analysis, and automated decision-making, the MAPE-K framework enhances system efficiency, optimizes resource utilization, and reduces operational costs, making it a crucial approach for self-adaptive computing environments.

- Adaptive Resource Scaling: The MAPE-K framework's continuous monitoring and analysis enable systems to adaptively scale resources in response to workload fluctuations, ensuring efficient resource utilization and maintaining performance levels.
- Improved Performance: By proactively detecting and addressing performance bottlenecks, the MAPE-K framework helps maintain system responsiveness and reliability, enhancing the overall user experience.
- Reduced Operational Costs: Through intelligent resource management, the MAPE-K framework minimizes unnecessary resource allocation, leading to cost savings without compromising service quality.

9.4.3 Challenges

Despite its benefits, implementing the MAPE-K framework presents challenges such as ensuring accurate predictive models, balancing resource allocation efficiently, and managing the computational complexity of real-time execution.

- Model Accuracy: The effectiveness of the MAPE-K framework relies heavily on the precision of its monitoring and analysis components. Inaccurate models can lead to suboptimal decisions, affecting system performance and resource utilization.
- Balancing Over- and Under-Provisioning: Achieving the right balance between over-provisioning (allocating more resources than necessary) and under-provisioning (allocating insufficient resources) is challenging. Both scenarios can have adverse effects, such as increased costs or degraded performance.
- Real-Time Execution Complexity: Implementing the MAPE-K framework in real-time systems requires handling substantial data volumes and making swift decisions, which can be complex and resource-intensive.

In conclusion, while the MAPE-K framework offers significant benefits in creating self-adaptive systems across various computing paradigms, addressing its inherent challenges is crucial for optimizing its application and realizing its full potential.

9.5 Performance Evaluation and Key Metrics

Performance evaluation in autoscaling systems relies on key metrics that measure the effectiveness of resource allocation and response efficiency. One of the most critical metrics is scaling response time, which determines how quickly the system reacts to workload fluctuations. A lower response time indicates a more efficient system capable of adapting to sudden demand changes without performance degradation. Other important metrics include resource utilization efficiency, which ensures that computing resources are optimally allocated without over- or under-provisioning, and cost efficiency, which evaluates whether the scaling strategy minimizes operational expenses while maintaining performance. Additionally, service availability and latency are crucial in cloud environments, as they directly impact user experience. A well-designed autoscaling system must balance these metrics to ensure optimal performance, cost-effectiveness, and system stability. Advanced techniques such as predictive analytics and reinforcement learning can further enhance autoscaling by improving response accuracy and reducing unnecessary scaling actions.

9.6 Real-World Implementation and Performance Evaluation Using the MAPE Framework

The MAPE (Monitor, Analyze, Plan, Execute) framework and its extension, MAPE-K (which incorporates a Knowledge component), play a crucial role in automating resource management in cloud environments. In real-world implementations, Kubernetes and Docker integrate MAPE to enable intelligent autoscaling, with Kubernetes utilizing the Horizontal Pod Autoscaler (HPA) and Vertical Pod Autoscaler (VPA) to dynamically adjust resources based on workload patterns. Docker enhances scalability by managing containerized applications efficiently, ensuring optimal performance with minimal resource wastage [128]. A case study on dynamic resource allocation in cloud computing demonstrates how MAPE-based autoscaling improves response time, system reliability, and cost efficiency by predicting demand and optimizing resource allocation. The MAPE-K framework, with its Knowledge component, further enhances decision-making by leveraging historical data and machine learning models for predictive autoscaling, thereby reducing latency and preventing performance degradation. Performance evaluations of these frameworks highlight improved scaling response times, efficient workload distribution, and reduced operational costs, proving their effectiveness in managing cloud-native applications. Future advancements, including AI-driven forecasting and adaptive learning mechanisms, can further refine these frameworks, making cloud infrastructure management more autonomous and resilient.

9.7 Real-Life Examples and Applications of MAPE and MAPE-K Frameworks

The MAPE and MAPE-K frameworks are widely used in real-world applications to enable autonomous and adaptive system management. These frameworks optimize cloud computing, edge computing, and AI-driven operations by dynamically adjusting resources based on real-time monitoring and predictive analytics.

- Cloud Resource Management in Kubernetes: Kubernetes employs the MAPE framework to manage cloud resources efficiently through its Horizontal Pod Autoscaler (HPA) and Vertical Pod Autoscaler (VPA) [125]. By continuously monitoring CPU and memory usage, Kubernetes analyzes workload patterns to determine optimal scaling actions. The HPA dynamically adjusts the number of pods based on real-time demand, ensuring seamless workload distribution, while the VPA modifies resource limits per pod, preventing under- or over-provisioning. This automated scaling mechanism optimizes cloud resource utilization, enhances system reliability, and reduces operational overhead, making Kubernetes a powerful tool for managing cloud-native applications.
- Self-Optimizing Edge Computing Systems: IBM Edge Application Manager utilizes the MAPE-K framework to enable intelligent resource management in edge computing environments where real-time decision-making is critical. By continuously monitoring edge devices, analyzing network conditions, and planning resource allocation, the system ensures optimal workload distribution. The execution phase dynamically adjusts resources, reducing latency and improving overall performance. This approach is particularly beneficial for IoT and 5G applications, where low-latency processing and efficient resource utilization are essential for maintaining seamless operations [129].
- AI-Driven Cloud Scaling with Reinforcement Learning: Google Cloud incorporates the MAPE-K framework to enable proactive autoscaling through predictive analytics and machine learning models [130]. By continuously monitoring cloud workloads, analyzing trends using AI-driven insights, and planning optimal scaling strategies, Google Cloud AutoML and Google Kubernetes Engine (GKE) dynamically adjust resources in real time. The execution phase ensures that workloads are efficiently distributed, preventing under-provisioning or over-provisioning, which enhances cost efficiency and system performance. This approach is particularly useful in cloud environments where workload fluctuations require intelligent and adaptive resource management.
- Autonomous Network Management in Telecommunications: Nokia AVA (Autonomous Virtual Assistant) leverages the MAPE-K framework for self-optimizing networks (SON), enabling real-time network optimization in telecommunications [131]. By continuously monitoring network traffic, analyzing congestion patterns, and planning reconfigurations, it dynamically adjusts network parameters to enhance performance and reliability. The execution phase ensures seamless adaptation to traffic fluctuations, reducing downtime and improving user experience. This AI-driven approach enhances operational efficiency, making networks more resilient and adaptive to changing demands in 5G and beyond.
- AI-Driven Self-Healing Systems: IBM Watson AIOps applies the MAPE-K framework to develop self-healing IT systems, ensuring autonomous anomaly detection and resolution in large-scale infrastructures [132]. By continuously monitoring system logs, analyzing patterns, and planning corrective measures, Watson AIOps anticipates failures and executes fixes without human intervention. This proactive approach minimizes operational overhead, enhances system

resilience, and ensures uninterrupted service availability, making it a crucial tool for IT operations and cloud management.

These real-world applications demonstrate how MAPE and MAPE-K frameworks enhance adaptive and autonomous system management across cloud computing, edge computing, networking, and AI-driven IT operations.

9.8 Advanced Research in Intelligent Autoscaling

Intelligent autoscaling is evolving with advanced machine learning models and hybrid strategies that optimize cloud resource management. The MV-Transformer Model has demonstrated state-of-the-art accuracy in workload prediction for Kubernetes-based applications, outperforming conventional LSTMs and BiLSTMs in forecasting demand fluctuations. Researchers are also exploring hybrid scaling strategies, such as integrating reinforcement learning (RL) with ant colony optimization (ACO) to enhance adaptability, ensuring efficient scaling decisions under dynamic workloads [133]. Additionally, meta-learning approaches are being introduced to improve generalization across different cloud environments, allowing models to adapt to unseen workload patterns with minimal retraining [134]. Future research directions focus on AI-driven self-healing autoscaling mechanisms, where autoscalers proactively detect performance anomalies, optimize resource allocation, and autonomously recover from failures. Another emerging area is energy-efficient scaling strategies, leveraging deep reinforcement learning (DRL) to balance performance and energy consumption, reducing carbon footprints in cloud data centers [135]. Moreover, the integration of digital twins in autoscaling frameworks is gaining traction, enabling real-time simulation of workload behavior to refine scaling decisions before actual deployment [136]. These advancements are driving the next generation of intelligent autoscaling, making cloud infrastructure more adaptive, autonomous, and sustainable.

9.9 Key Takeaways

- The MAPE (Monitor, Analyze, Plan, Execute) framework automates decision-making for resource management, enabling intelligent autoscaling in cloud environments.
- MAPE-K extends MAPE by incorporating a Knowledge component, allowing predictive scaling based on historical data and AI-driven insights.
- Proactive autoscaling strategies are more efficient than reactive methods as they anticipate workload surges, reducing latency and preventing system overload.
- Machine learning models integrated with proactive autoscaling enhance performance optimization and cost efficiency in dynamic cloud infrastructures.

- Kubernetes' Horizontal Pod Autoscaler (HPA) and Vertical Pod Autoscaler (VPA) utilize MAPE principles to dynamically adjust resources, ensuring scalability and stability.
- Docker containerization enhances workload distribution, minimizes resource wastage, and improves system resilience in cloud-native environments.
- MV-Transformer-based models provide state-of-the-art accuracy in workload prediction, outperforming conventional forecasting techniques for precise resource allocation.
- The future of intelligent autoscaling involves AI-driven self-healing mechanisms, hybrid optimization strategies, and energy-efficient scaling techniques for adaptive cloud operations.

9.10 Summary

The chapter delves into the MAPE (Monitor, Analyze, Plan, Execute) framework, which plays a crucial role in intelligent autoscaling within cloud environments. It begins by introducing the fundamental components of MAPE, explaining how it automates resource management by continuously monitoring system performance, analyzing workload patterns, planning scaling decisions, and executing them dynamically. This automation ensures that cloud resources are allocated efficiently, reducing both latency and operational costs. A key distinction is made between reactive and proactive autoscaling. Reactive autoscaling responds to real-time workload fluctuations, adjusting resources as needed to maintain system performance. In contrast, proactive autoscaling takes a more sophisticated approach by leveraging predictive models to anticipate future demand. By forecasting resource requirements, proactive autoscaling minimizes the risk of under-provisioning, which can degrade performance, or over-provisioning, which leads to unnecessary costs. Building upon the traditional MAPE model, the MAPE-K framework incorporates a Knowledge component, allowing for more intelligent decision-making. This enhanced version utilizes historical data, machine learning, and AI-driven analytics to optimize autoscaling strategies. The chapter explores various applications of MAPE-K, discussing its advantages and the challenges associated with its implementation in large-scale cloud infrastructures. To assess the effectiveness of these frameworks, the chapter includes a performance evaluation section, focusing on key metrics such as scaling response time. The ability to efficiently handle workload fluctuations is a critical measure of autoscaling success. The discussion highlights real-world implementations, particularly how Kubernetes and Docker integrate MAPE-based autoscaling techniques. Kubernetes employs both Horizontal Pod Autoscaler (HPA) and Vertical Pod Autoscaler (VPA) to dynamically scale applications based on workload patterns, while Docker enhances resource efficiency by managing containerized applications. These implementations demonstrate how MAPE frameworks optimize cloud-native applications, ensuring adaptability and performance stability. The chapter also explores advanced research in intelligent autoscaling, showcasing innovations that push the boundaries of traditional scaling mechanisms. The MV-Transformer model is highlighted as a state-of-the-art approach that enhances workload prediction accuracy, improving autoscaling efficiency in Kubernetes. Additionally, hybrid scaling strategies that combine reinforcement learning and ant colony optimization offer more adaptable and responsive autoscaling solutions. Future research in this field aims to develop AI-driven self-healing mechanisms that autonomously detect and resolve performance issues, as well as energy-efficient scaling strategies that minimize environmental impact while maintaining optimal system performance. In conclusion, the chapter underscores the significance of MAPE-driven autoscaling in modern cloud computing. It emphasizes that proactive autoscaling outperforms reactive methods by enabling more efficient resource allocation, reducing operational costs, and enhancing system resilience. The integration of machine learning and AI further strengthens the capabilities of MAPE and MAPE-K, paving the way for more autonomous and intelligent cloud infrastructure management.

9.11 Review Questions

- What are the four core components of the MAPE framework, and how do they interact in autoscaling?
- How does the MAPE framework enhance reactive autoscaling in Kubernetes?
- What are the main advantages of using reactive autoscaling with MAPE?
- What limitations does reactive autoscaling face when implemented via MAPE?
- How does proactive autoscaling differ from reactive when using the MAPE framework?
- What is the role of the Knowledge (K) component in MAPE-K, and how does it improve decision-making?
- In what scenarios is MAPE-K more effective than standard MAPE for autoscaling?
- What are the key advantages of using MAPE-K for predictive autoscaling?
- What challenges arise when implementing MAPE-K in dynamic cloud environments?
- Which performance metrics are most critical when evaluating MAPE-based autoscaling systems?
- How can real-world implementations validate the effectiveness of the MAPE framework?
- What are some real-life examples where MAPE or MAPE-K improved autoscaling efficiency?
- How do cloud providers leverage MAPE-K for intelligent resource management?
- What cutting-edge research is being conducted to enhance MAPE-based autoscaling?
- How might AI and machine learning further optimize the MAPE-K loop in future systems?

Chapter 10
Proactive Autoscaling for Cloud-Native Applications Using Predictive Resource Allocation

Abstract This chapter focuses on proactive autoscaling strategies for cloud-native applications through predictive resource allocation. It begins by examining the challenges of cloud-native architectures and the limitations of reactive autoscaling in responding to dynamic workloads. Proactive autoscaling is presented as a solution, leveraging predictive models to anticipate demand and optimize resource allocation. The InformerAutoScale framework is introduced, including its architectural overview, key algorithms, and core innovations that enable efficient scaling decisions. The methodology section details the data processing pipeline, Kubernetes integration, and execution layer for real-time implementation. Performance evaluation demonstrates the framework's effectiveness in maintaining low latency, high resource utilization, and cost efficiency. The chapter concludes with a discussion of insights, highlighting the practical advantages of predictive autoscaling for both cloud and edge environments. By integrating predictive analytics with Kubernetes orchestration, the approach ensures reliable and intelligent resource management in dynamic application scenarios.

Keywords Proactive autoscaling · Predictive resource allocation · Cloud-native applications · Kubernetes · InformerAutoScale framework

Modern cloud-native applications are specifically engineered to operate in distributed cloud environments, taking full advantage of cloud computing's elastic nature [137, 138]. These applications are designed with microservices, containerization, and stateless architectures that allow them to scale horizontally across multiple cloud nodes with minimal friction. Elasticity [138], one of the key features of cloud computing, allows these applications to dynamically acquire or release resources based on demand, which is critical in today's fast-paced and variable user environments. As organizations increasingly adopt containerized architectures, efficient resource management has become paramount [139]. The shift toward Docker and Kubernetes has introduced agility in deployment and scalability, but it also demands precise control over CPU, memory, and network resources. Inefficient

B. Kumar et al., *Modern Kubernetes: From Core Concepts to Intelligent Autoscaling for Cloud Applications*, Studies in Autonomic, Data-driven and Industrial Computing, https://doi.org/10.1007/978-3-032-12972-7_10

resource allocation leads to either service degradation or unnecessary operational costs. Traditional reactive scaling methods, which respond to real-time system metrics, often struggle with latency issues and inefficient resource utilization during unpredictable workload fluctuations [140]. Reactive scaling decisions, triggered only after system metrics like CPU utilization cross a threshold, often arrive too late. This delay can result in missed service level objectives (SLOs), and during volatile workloads, these methods frequently alternate between scaling up and down, creating instability. This chapter presents an innovative approach to cloud resource management through InformerAutoScale, a proactive autoscaling framework that combines predictive analytics with container orchestration. InformerAutoScale introduces a forward-looking methodology that leverages machine learning-based forecasting to anticipate workload trends [141]. It integrates this intelligence into the Kubernetes orchestration workflow, allowing for seamless and timely resource scaling. Unlike conventional methods, our solution anticipates resource demands before they occur, enabling preemptive scaling decisions that maintain optimal performance while minimizing both under-provisioning and over-provisioning scenarios. By predicting workload changes ahead of time, InformerAutoScale ensures that sufficient resources are available before demand surges. This not only protects performance but also avoids wasting resources during idle periods, optimizing both cost and reliability.

10.1 Background and Motivation

The growing complexity and scale of cloud-native applications demand intelligent resource management strategies. This section explores the key challenges in containerized environments and the limitations of current autoscaling approaches, underscoring the need for proactive solutions.

10.1.1 Cloud-Native Architecture Challenges

Cloud-native architecture enables scalable and resilient application deployment, but it also introduces new complexities in resource management. As workloads grow more dynamic and unpredictable, addressing these architectural challenges becomes crucial for maintaining performance and cost efficiency [139].

- Dynamic Workload Variability: Cloud-native applications, especially those serving global user bases or handling real-time data streams, experience sudden and irregular spikes or drops in demand. This variability makes it difficult to rely on static provisioning or delayed response systems.
- Resource Provisioning Dilemmas: Reactive systems typically rely on predefined thresholds, which may not suit all workloads. If the threshold is too low, the system scales too late (under-provisioning), affecting performance. If too

high, it allocates more resources than necessary (over-provisioning), increasing operational costs.
- Scaling Latency: Even with Kubernetes HPA, scaling involves detecting a metric breach, triggering a decision, and launching new pods—a process that can take several minutes. This delay is significant during rapid workload surges, where every second of delay can impact user experience or service availability.

These issues highlight the limitations of conventional scaling strategies and the necessity for smarter, faster, and predictive mechanisms in managing cloud-native environments.

10.1.2 Reactive Autoscaling

Reactive autoscaling remains a widely implemented strategy due to its simplicity and integration with container orchestration platforms like Kubernetes [137]. However, its reliance on real-time system feedback introduces critical limitations in dynamic and latency-sensitive environments.

- Responds to real-time metrics (CPU, memory usage): These approaches continuously monitor resource usage and initiate scaling only after thresholds are crossed. While simple and widely adopted, they are inherently limited by their dependence on past or current state metrics.
- These approaches continuously monitor resource usage and initiate scaling only after thresholds are crossed. While simple and widely adopted, they are inherently limited by their dependence on past or current state metrics.
- Suffers from inherent latency in scaling operations: Since decisions are made after stress is already detected, the system is always playing catch-up. This reactive nature often fails to meet latency-sensitive application needs.
- Often creates resource thrashing during volatile workloads: Rapid up-and-down scaling due to frequent threshold crossing leads to resource instability. Pods may be added and removed repeatedly within short timeframes, wasting compute cycles and affecting performance consistency.

Despite being easy to implement, reactive methods fall short in addressing the real-time and predictive needs of modern cloud-native workloads, often resulting in performance bottlenecks and operational inefficiencies.

10.1.3 Proactive Autoscaling

Proactive autoscaling techniques aim to overcome the latency of reactive systems by forecasting future resource demands. While promising in theory, current implementations face significant barriers in real-world, production-scale environments.

- Uses historical data to predict future demands: These methods rely on time-series analysis and machine learning models trained on past resource usage to forecast future needs, enabling scaling ahead of time [142].
- Existing predictive models (LSTM, Transformer) struggle with long-sequence forecasting: LSTM and Transformer models, though powerful, often encounter challenges when dealing with long historical sequences. They may require significant computational resources and time for both training and inference, which hinders real-time application.
- Often computationally expensive for real-time decision making: Deploying these models in a production autoscaling loop can add overhead, negating their benefits. Frequent retraining, high memory consumption, and longer inference times can delay timely scale-out decisions.

Despite their predictive potential, current proactive methods are often too complex or slow to support dynamic workloads efficiently. These challenges emphasize the need for a lightweight, accurate, and latency-aware autoscaling framework—one that this chapter introduces through the InformerAutoScale solution.

10.2 Autoscaling Models

This section provides an in-depth evaluation of traditional and modern autoscaling models, comparing their strengths and weaknesses in both reactive and proactive contexts. Understanding these models' capabilities is crucial for selecting or designing an effective autoscaling framework for cloud-native applications.

- ARIMA (AutoRegressive Integrated Moving Average): ARIMA is a traditional time-series forecasting model valued for its simplicity and interpretability, particularly effective with stationary data and short-term dependencies. While it can capture seasonal or periodic trends when properly configured, ARIMA's linear assumptions limit its ability to model the complex, nonlinear workload patterns typical of cloud environments. This results in poor performance when dealing with long sequences or sudden workload spikes, such as viral traffic surges. Additionally, ARIMA requires manual parameter tuning, which complicates its use in automated, large-scale cloud systems, making it less suitable for highly dynamic and unpredictable workloads.
- LSTM (Long Short-Term Memory): LSTM networks offer significant improvements over traditional models by effectively capturing complex temporal dependencies in sequential data through their memory cells, which retain information over extended periods. This capability makes them well-suited to understanding workload fluctuations, particularly for proactive autoscaling where forecasting future demand is critical. However, LSTMs face challenges in reactive scenarios due to high inference latency that limits quick responsiveness. Their sequential processing also restricts parallelization, leading to difficulties with very long sequences and scalability issues in real-time production environments,

compounded by their substantial computational cost and lengthy training requirements.

- Bi-LSTM (Bidirectional LSTM): Bi-LSTMs build upon standard LSTMs by processing data sequences bidirectionally, allowing them to capture both past and future context simultaneously. This enhanced understanding of workload trends can improve reactive scaling decisions by providing more comprehensive insights into system behavior. However, this comes at the cost of roughly doubling the computational requirements compared to regular LSTMs, which can reduce real-time responsiveness. In proactive autoscaling scenarios, Bi-LSTMs tend to deliver better accuracy, particularly for workloads exhibiting periodic or cyclical patterns. Nonetheless, their inherently slow training and inference times pose challenges for large-scale or time-sensitive environments where rapid prediction and scalability are critical.
- Transformer: Transformers leverage self-attention mechanisms that allow them to process sequences in parallel, resulting in significantly faster inference times compared to LSTM-based models. This parallelism makes them well-suited for reactive autoscaling scenarios that require rapid decision-making. However, the quadratic growth in computational complexity with sequence length ($O(n^2)$) can hinder their performance on very long input sequences. In proactive autoscaling, Transformers excel by effectively capturing global workload patterns, which improves the accuracy of long-range forecasts. Despite these advantages, their high memory demands and need for large training datasets pose challenges, especially in resource-constrained environments, potentially limiting their practical deployment.
- Informer: Informer is a recent advancement designed specifically to address the limitations of Transformers in time-series forecasting. It introduces ProbSparse attention, which reduces the complexity from ($O(n^2)$) to $O(n \log n)$, making it more scalable for long sequences common in cloud workloads. The distilled decoder further optimizes the forecasting by condensing encoder outputs, enabling efficient multi-step prediction. Additionally, Informer employs Time2Vec embeddings that enhance temporal pattern recognition beyond what positional encodings offer. These innovations make Informer highly suited for proactive autoscaling, achieving impressive prediction accuracy (94.2%) for 5-minute forecasting horizons.

This Table 10.1 summarizes the suitability and key limitations of various autoscaling models across proactive and reactive scenarios. It highlights how each model balances prediction accuracy with computational costs, guiding the selection of appropriate methods for different cloud workload challenges.

By comparing these models, it becomes evident that while traditional models like ARIMA offer simplicity and interpretability, they fall short in capturing the complexity of cloud workloads. Deep learning models such as LSTM and Transformer improve forecasting accuracy but introduce challenges related to computation time and resource consumption. Informer, with its novel attention mechanism and efficiency improvements, represents a promising direction for

Table 10.1 Summary table of model suitability and limitations

Model	Proactive suitability	Reactive suitability	Key limitation
ARIMA	Low	Medium	Linear assumptions
LSTM	Medium	Low	Sequential bottleneck
Bi-LSTM	Medium	Low	High computational cost
Transformer	High	Medium	Memory overhead
Informer	High	Low	Requires periodic retraining

building scalable and accurate proactive autoscaling systems, balancing prediction performance with computational feasibility.

10.3 The InformerAutoScale Framework

The InformerAutoScale Framework is developed to tackle the increasingly complex challenge of managing dynamic and unpredictable workloads in cloud-native environments, particularly those orchestrated using platforms like Kubernetes. Traditional autoscaling approaches, such as threshold-based or reactive scaling mechanisms, respond only after resource utilization crosses certain predefined limits. While simple, this reactive nature leads to delays, instability, and inefficient use of computational resources. InformerAutoScale overcomes these shortcomings by leveraging a proactive, machine learning-driven methodology [142].

- Advanced Forecasting At the core of InformerAutoScale lies a powerful forecasting engine that predicts future resource demands based on historical workload patterns. This component utilizes state-of-the-art time-series prediction models, such as the Informer architecture—a deep learning model optimized for long sequence forecasting with reduced computational complexity. By analyzing trends in CPU usage, memory consumption, and network traffic, the framework anticipates scaling needs before they occur. This enables preemptive provisioning or deprovisioning of pods, reducing latency and preventing bottlenecks.
- Adaptive Control Logic Unlike rigid rule-based systems, InformerAutoScale includes an adaptive control layer that dynamically adjusts scaling thresholds and policies based on the predicted workload and current system state. This allows it to fine-tune the scale-in and scale-out operations, avoiding problems such as resource thrashing (frequent scaling up and down) or underutilization. The control logic incorporates factors such as forecast confidence intervals, pod startup time, and SLA constraints to make intelligent scaling decisions.
- Seamless Orchestration Integration The framework is natively integrated with Kubernetes through custom controllers and metrics adapters. It monitors system performance using the Kubernetes metrics server and applies scaling actions via the Horizontal Pod Autoscaler (HPA) or a custom scaling engine. This tight integration ensures that the predictive scaling decisions translate into real-time

actions without disrupting application performance. Additionally, it leverages Kubernetes' declarative configuration and reconciliation loop to maintain the desired system state.

- Proactive Resource Scaling InformerAutoScale adopts a proactive scaling strategy, meaning it allocates resources before they are actually needed—unlike reactive systems that lag behind demand. This approach improves user experience during sudden load spikes (e.g., flash sales, live events, or traffic bursts) by ensuring that the infrastructure is prepared ahead of time. It reduces cold start latency and improves SLA compliance, ensuring consistent performance across varying workloads.
- Deployment Stability and Efficiency A major goal of the framework is to preserve system stability while achieving high resource efficiency. By forecasting and scaling gradually based on demand predictions, it avoids sudden large-scale changes that could destabilize the system or introduce service interruptions. Moreover, by eliminating excessive over-provisioning, the framework reduces unnecessary cloud resource consumption—translating into significant cost savings for organizations.
- Scalability and Extensibility The architecture of InformerAutoScale is designed to be modular and extensible, allowing integration with other forecasting models, control algorithms, or cloud platforms beyond Kubernetes. This makes it suitable for hybrid cloud and edge computing environments, where decentralized workload patterns add further complexity to resource allocation.

10.3.1 Architectural Overview

The architecture of *InformerAutoScale* is structured into three integral components that function cohesively to deliver intelligent, proactive autoscaling capabilities in cloud-native environments [143]. Each component addresses a specific stage in the autoscaling pipeline—from workload forecasting to dynamic resource provisioning within Kubernetes.

- **Predictive Engine:**
  At the heart of the framework is the Informer model, an advanced time-series forecasting neural network optimized for long-sequence predictions. Unlike traditional LSTMs or GRUs, the Informer leverages ProbSparse self-attention mechanisms that significantly reduce computational complexity while preserving global temporal dependencies. This efficiency enables the engine to handle high-dimensional telemetry data, such as CPU usage, memory consumption, and network traffic, across extended time horizons. The engine receives historical resource utilization metrics at regular intervals and processes them to produce multistep-ahead forecasts. These forecasts capture both cyclical patterns (e.g., daily user access trends) and sudden spikes (e.g., flash crowds or microservice fan-outs). The output is a future workload matrix that serves as the primary

input for scaling decisions, thereby eliminating latency introduced by threshold breaches in reactive systems.

- **Adaptation Manager:**
  Sitting at the decision-making core, the Adaptation Manager executes a modified MAPE (Monitor–Analyze–Plan–Execute) loop enhanced with predictive capabilities. It continuously ingests live operational metrics and compares them against the predictive engine's output.
  In the *Monitor* phase, real-time data such as pod CPU load, memory footprint, and request queue length are collected from Kubernetes. During *Analyze*, these metrics are aligned with the predicted workload trajectory to identify future capacity gaps. The *Plan* phase then determines the optimal number of pods or replicas needed to meet the forecasted load while minimizing over-provisioning. Finally, the *Execute* phase invokes autoscaling commands through Kubernetes APIs to add or remove pods dynamically.
  By incorporating forecast awareness into the control loop, the adaptation manager proactively prepares the infrastructure ahead of time rather than reacting to breaches. This minimizes resource thrashing, cold starts, and SLA violations.
- **Kubernetes Integration:**
  To operationalize predictions and planned scaling actions, the framework integrates tightly with Kubernetes, the de facto orchestrator in cloud-native deployments. It employs a custom configuration of the Horizontal Pod Autoscaler (HPA) that supports external metrics. Instead of relying solely on built-in metrics such as CPU or memory usage, this setup enables HPA to respond to predicted workload signals exported from the predictive engine.
  The system registers a new metric type—`predicted_request_rate`—to Kubernetes via a metrics adapter. This adapter acts as a bridge, translating the forecasting engine's output into a format compatible with Kubernetes HPA. The following configuration snippet illustrates how an external metric is referenced:

```
metrics:
- type: External
  external:
    metric:
      name: predicted_request_rate
    target:
      type: Value
      value: 1000
```

Listing 10.1 Kubernetes HPA external metrics configuration

The YAML snippet represents the configuration of a Kubernetes Horizontal Pod Autoscaler (HPA) that utilizes external metrics to make scaling decisions. Specifically, it configures the HPA to monitor a custom metric named `predicted_request_rate`, which is provided by an external source—typically a forecasting model like the Informer engine in the *InformerAutoScale* framework. The `type: External` directive signals that the metric is not a built-in resource metric (such as CPU or memory) but instead comes from an external system, such as a Prometheus adapter or custom metrics API. The `target` section specifies the desired threshold value for the metric, in this case `1000`, indicating that when the predicted request rate reaches or exceeds this value, the HPA will trigger scaling actions—such as increasing the number of pods. This proactive approach allows Kubernetes to react not to current usage but to forecasted demand, helping prevent performance degradation during sudden traffic spikes and enhancing overall resource efficiency.

This integration empowers Kubernetes to trigger autoscaling events based not on current stress, but on anticipated future demand, thereby improving pod startup readiness, cost efficiency, and service stability. Moreover, it allows the framework to remain modular and extensible, supporting future enhancements like multi-metric scaling or cross-cluster federation.

10.3.2 Key Algorithms

The autoscaling logic of the *InformerAutoScale* framework relies on carefully crafted algorithms to ensure both the accuracy of scaling decisions and the overall stability of the system. These algorithms are designed to anticipate future workload demands and execute scaling actions in a way that minimizes disruptions and optimizes resource usage.

- **Proactive Scaling Decision:**
  At the heart of the autoscaling framework is the proactive scaling algorithm, which determines the number of pods needed to handle forecasted workloads before they occur. This algorithm takes the predicted request rate—generated by the forecasting engine—and divides it by the known capacity of a single pod to calculate the minimum number of pods required to maintain performance. To ensure that the system operates within safe and efficient boundaries, the calculated number of pods is then constrained by predefined minimum and maximum pod limits. This prevents extreme scaling that could either overwhelm resources or lead to excessive costs. By relying on forecasted data rather than reactive signals alone, this approach reduces latency spikes and ensures that sufficient resources are available in anticipation of demand.

```
def scale_pods(forecast, current_pods):
    desired_pods = ceil(forecast / POD_CAPACITY)
    return clamp(desired_pods, MIN_PODS, MAX_PODS)
```

Listing 10.2 Proactive scaling function

- **Resource Removal Strategy (RRS):**
  Autoscaling systems can suffer from a phenomenon called resource thrashing, where rapid and frequent scaling up and down causes instability, inefficiencies,

and potential service degradation. To address this, the framework implements a gradual resource removal strategy that smooths out scaling down actions. Rather than instantly terminating all pods beyond the desired count, the algorithm removes only 60% of the excess pods in each cycle. This staged removal avoids sudden drops in capacity that could impact performance and allows the system to stabilize over time. The RRS effectively balances resource optimization with system reliability, minimizing oscillations in pod counts while keeping costs under control.

Together, these algorithms form a robust mechanism that empowers *Informer-AutoScale* to proactively manage resources in cloud-native environments, ensuring responsiveness to workload changes while maintaining stability and cost efficiency.

10.3.3 Core Innovations

The InformerAutoScale framework stands out due to several novel advancements that improve both the accuracy of workload forecasting and the effectiveness of resource scaling in cloud-native environments. These innovations can be categorized into three primary areas: enhanced forecasting capability, dynamic resource adjustment, and stabilization mechanisms.

- **Enhanced Forecasting Capability:**
  At the heart of InformerAutoScale's predictive power is the ProbSparse self-attention mechanism. Traditional self-attention models require computational resources proportional to the square of the input length ($O(n^2)$), which becomes impractical for long time-series datasets common in workload histories. ProbSparse attention cleverly reduces this complexity to near-linear ($O(n \log n)$) by focusing attention on the most significant parts of the sequence, making it highly efficient and scalable for extended forecasting horizons. This efficiency is critical when anticipating future resource demands in dynamic cloud workloads [144].
  In addition, the model incorporates Time2Vec embeddings instead of classical positional encodings. Time2Vec captures both periodic and non-periodic temporal features of the data, enabling the model to better understand and predict complex seasonal patterns and trends in workload behavior. This enhanced temporal representation leads to more accurate and robust predictions.
  Moreover, the use of a distilled decoder architecture condenses the output of the encoder. By summarizing essential information from long input sequences, this approach reduces the computational and memory demands during decoding, allowing the system to handle longer forecast horizons without overwhelming resource use.
- **Dynamic Resource Adjustment:**
  The framework translates forecasted workloads into precise scaling actions using a straightforward yet effective formula. By dividing the predicted load (e.g., incoming request rate) by the capacity of a single pod (i.e., the maximum

requests a pod can handle), the system calculates the ideal number of pods needed. This count is then bounded by predefined minimum and maximum limits to avoid unrealistic scaling decisions—preventing under-provisioning that risks performance degradation, and over-provisioning that wastes resources.

```
def calculate_pod_requirements(forecasted_load, current_pods)
    :
    pod_capacity = 1000  # requests per pod
    desired_pods = ceil(forecasted_load / pod_capacity)
    return clamp(desired_pods, MIN_PODS, MAX_PODS)
```

Listing 10.3 Pod requirement calculation

This mechanism is implemented in a concise function that determines the number of pods to deploy, ensuring scaling decisions are both grounded in predictive analytics and constrained within operational limits. This clear formula helps automate scaling decisions while respecting real-world infrastructure boundaries.

- **Stabilization Mechanisms:**
  To avoid common pitfalls in autoscaling such as rapid oscillations or "thrashing," the framework incorporates multiple stabilization techniques. The Container Deployment Target (CDT) introduces a deliberate 60-second delay between consecutive scaling actions, effectively spacing them out. This buffer time prevents the system from reacting too frequently to transient workload spikes, which could otherwise cause instability.
  Additionally, the Resource Removal Strategy (RRS) moderates scale-down operations by gradually decommissioning pods. Instead of abruptly removing all surplus pods, only a portion (e.g., 60%) of excess resources is removed during each scaling cycle. This controlled reduction smooths resource allocation changes, maintaining service stability and preventing sudden drops in capacity that might negatively impact application performance.

Together, these innovations create a cohesive and robust autoscaling framework that leverages cutting-edge forecasting methods combined with practical control and deployment strategies. InformerAutoScale thus effectively manages the challenges of dynamic cloud workloads, delivering scalable, efficient, and stable performance in production environments.

10.4 Methodology Used

This section outlines the complete implementation workflow of our autoscaling system, which combines a deep-learning-based prediction model with container orchestration via Kubernetes. The methodology comprises two major components: the data processing pipeline and the Kubernetes-based deployment architecture. Each of these is discussed in detail below.

10.4.1 Data Processing Pipeline

The predictive capability of our system relies on robust data preprocessing and model training. The pipeline is designed to ensure high data quality, meaningful feature extraction, and optimal model performance for forecasting workload demand.

- Data Collection: We sourced real-world data from two prominent datasets to simulate diverse workload environments. First, we utilized the NASA HTTP web server logs [145], which consist of approximately 3.4 million request entries. These logs capture user interactions with web content, including timestamps, requested resources, and response codes—critical for modeling user-driven workload patterns.
  Second, we leveraged the Google Cluster Traces dataset, which records the activity of over 12,500 machines in a production-level cluster over 29 days. This dataset includes granular information on task submissions, scheduling, and resource utilization, making it ideal for understanding infrastructure behavior under varying load conditions. Combining these datasets provides both user-level and infrastructure-level perspectives for model generalization.
  The inclusion of the Google Cluster Traces enables the system to learn how backend resource utilization correlates with user requests, helping it predict workload impacts on compute, memory, and network resources in a realistic environment.
- Feature Engineering: To enhance the model's ability to detect patterns and trends, we engineered several features from the raw data. Temporal features were derived by decomposing timestamps into components such as hour of the day, day of the week, and day-night cycles. These patterns often correlate with user behavior [146] and system load, which helps the model anticipate cyclical spikes or troughs.
  In addition to temporal dimensions, we extracted resource utilization metrics such as CPU, memory, disk I/O, and network throughput. These metrics provide insight into current system stress and are essential [147] for aligning workload predictions with infrastructure constraints.
  We also computed request rate derivatives—measuring how fast request rates are increasing or decreasing over short intervals. These first and second-order statistics help the model capture bursty behavior and accelerate scaling decisions in dynamic environments [148]. This derivative information enables the model to respond more quickly to sudden workload surges or drops by anticipating acceleration or deceleration trends in request rates, leading to more proactive autoscaling.
- Model Training: The processed dataset was split using a 70/30 train-test ratio to preserve temporal ordering and avoid data leakage. For rigorous model validation, we employed time-series cross-validation instead of random shuffling, which respects temporal dependencies and better reflects real-world performance [149].

We trained the Informer-based forecasting model using the Adam optimizer with an initial learning rate of 0.001 [150]. The model's learning rate was tuned over multiple runs to ensure convergence and stability. To prevent overfitting, we implemented early stopping based on validation Mean Absolute Error (MAE). Training halted when the validation error plateaued for a defined number of epochs, ensuring the model generalizes well to unseen workloads. This training strategy balances model complexity and generalization, ensuring that the forecasting model remains robust and accurate when applied to real-time scaling decisions.

10.4.2 Kubernetes Integration

The predictive autoscaling model was integrated into a Kubernetes environment to facilitate real-time container orchestration [151]. This integration was achieved through a multi-layered deployment architecture comprising monitoring, decision-making, and execution layers.

- Monitoring Layer: To feed real-time data into the scaling logic, we established a robust monitoring layer. This layer uses `cAdvisor`, a lightweight container advisor tool, to collect low-level resource usage statistics directly from Docker containers. It tracks metrics such as CPU, memory, and I/O, which are vital for dynamic scaling decisions. By collecting these detailed container-level statistics, `cAdvisor` ensures the autoscaler has timely and precise insights into the resource consumption patterns of each pod, which is critical for informed scaling.
  In addition to `cAdvisor`, we developed custom metric exporters tailored to specific applications. These exporters extract application-level metrics like request rates, latency, and error rates, which are not captured by default tools. These metrics are exported in a Prometheus-compatible format to be consumed by the autoscaler. This application-specific monitoring allows the autoscaler to factor in business-relevant indicators (e.g., request throughput or error frequency), thus enabling more nuanced and effective scaling beyond just infrastructure metrics.
- Decision Engine: The decision engine governs the logic behind when and how scaling should occur. It is built on top of Kubernetes' Horizontal Pod Autoscaler (HPA) with support for external metrics. The HPA is configured to monitor a custom metric named `predicted_request_rate`, which is generated by our forecasting model.
  The autoscaler dynamically adjusts the number of pods based on this forecasted value. A threshold target value of 1000 requests per pod is specified. The scaling configuration allows for a minimum of 2 pods and a maximum of 20 pods to maintain resource efficiency while avoiding overload.

```yaml
apiVersion: autoscaling/v2
kind: HorizontalPodAutoscaler
metadata:
  name: informer-autoscaler
spec:
  scaleTargetRef:
    apiVersion: apps/v1
    kind: Deployment
    name: web-application
  minReplicas: 2
  maxReplicas: 20
  metrics:
  - type: External
    external:
      metric:
        name: predicted_request_rate
      target:
        type: Value
        value: 1000
```

Listing 10.4 Kubernetes HPA configuration

The `apiVersion: autoscaling/v2` specifies that this Horizontal Pod Autoscaler (HPA) configuration uses Kubernetes API version `v2`. This version supports advanced autoscaling features, including the ability to scale based on multiple metrics and external custom metrics, which are essential for implementing predictive autoscaling strategies beyond basic CPU or memory utilization.

The `kind: HorizontalPodAutoscaler` declares that this resource manages automatic horizontal scaling of pods within a Kubernetes cluster [152]. The HPA monitors specified metrics and dynamically adjusts the number of pod replicas in a deployment to maintain optimal performance and efficient resource usage, reducing manual intervention.

Under `metadata`, the HPA is given the name `informer-autoscaler`, which uniquely identifies it within the Kubernetes environment. This helps administrators and monitoring tools recognize and manage the autoscaler configuration associated with the deployment.

Within the `spec` section, the key element `scaleTargetRef` points to the Kubernetes Deployment named `web-application`. This tells the HPA exactly which deployment it should control, so it can increase or decrease the number of pods in this deployment based on observed metrics.

The fields `minReplicas: 2` and `maxReplicas: 20` define the scaling boundaries. The autoscaler will never scale below two pods, ensuring a minimum service availability and baseline capacity. Conversely, it will not scale above twenty pods, which helps control resource usage and prevents over-provisioning that could lead to unnecessary costs or cluster overload.

The `metrics` section is where the power of predictive autoscaling is configured. Here, the autoscaler monitors an external custom metric instead of the default CPU or memory metrics. Specifically, it watches the `predicted_request_rate`, which is a forecasted value generated by a deep-learning model that predicts future incoming request rates for the application.

By setting the target type as `Value` with a threshold of `1000`, the autoscaler expects each pod to handle approximately 1000 requests per unit time. It calculates the desired number of pods by dividing the forecasted request rate by this target. For example, if the prediction indicates 8000 requests, the HPA will scale up the deployment to around 8 pods.

This approach enables **proactive scaling** based on predicted future workload rather than reacting to current resource usage or load. By anticipating spikes or drops in demand, the autoscaler can adjust resources in advance, leading to reduced latency during peak periods and avoiding unnecessary resource consumption during low usage times.

Overall, this configuration allows the Kubernetes cluster to maintain a balance between performance and efficiency, ensuring that the application can handle dynamic workloads smoothly while optimizing infrastructure costs and responsiveness.

10.4.3 Execution Layer

The execution layer represents the crucial final stage in our autoscaling deployment architecture. Its primary role is to translate the scaling decisions generated by the predictive model into concrete actions within the Kubernetes cluster. To achieve this, we developed a custom controller implemented in Python, which interacts directly with the Kubernetes API. This controller acts as a bridge between the forecasting model and Kubernetes' native autoscaling mechanisms. It continuously retrieves the forecasted workload metrics from the prediction model and forwards these to the Horizontal Pod Autoscaler (HPA), thereby enabling the system to adjust resources based on anticipated demand rather than solely on current utilization.

One of the key challenges in autoscaling is avoiding abrupt or drastic changes in resource allocation, which can destabilize running services and degrade user experience. To mitigate this risk, we designed and integrated gradual scaling policies within the controller logic. These policies impose limits on how quickly the number of pods can increase or decrease in any single control interval. For instance, when scaling down, the controller restricts pod removal to no more than 60% of the surplus pods per cycle. This throttling mechanism ensures that scaling happens smoothly over multiple intervals, allowing the system to maintain stability, avoid sudden workload shocks, and better handle transient forecasting inaccuracies or metric fluctuations.

By combining this careful execution strategy with data-driven forecasting, our methodology ensures that autoscaling not only anticipates future demand but also implements scaling actions in a controlled and resilient manner. This integrated approach maintains high service availability, improves system responsiveness, and optimizes resource utilization even under highly dynamic and unpredictable workload conditions.

10.5 Performance Evaluation

To thoroughly assess the effectiveness of the InformerAutoScale framework, a comprehensive experimental setup was established. The testing environment utilized Kubernetes version 1.28, running on Docker Desktop with the Windows Subsystem for Linux 2 (WSL2) backend. The host system was equipped with 8 GB of RAM and 2 virtual CPUs, simulating a modest yet realistic edge or cloud environment. This configuration provides a suitable platform for benchmarking lightweight autoscaling solutions commonly deployed in constrained or distributed settings, such as edge data centers or small-scale cloud clusters [153].

The InformerAutoScale framework was evaluated against several widely recognized baselines to ensure a fair and comprehensive comparative analysis. These baselines included Long Short-Term Memory (LSTM) networks and Transformer-based models, both popular deep learning architectures for time-series forecasting, as well as the native Kubernetes Horizontal Pod Autoscaler (HPA), which represents the standard reactive autoscaling approach. This evaluation strategy allowed us to rigorously compare the predictive and operational performance of InformerAutoScale against both traditional machine learning models and Kubernetes' default scaling mechanisms [142, 154].

The evaluation focused on several key performance metrics critical to autoscaling efficiency and service reliability. These included the maximum pod count, under-provisioning rates, over-provisioning rates, and adherence to the response time Service Level Agreement (SLA). InformerAutoScale demonstrated a remarkable reduction in the maximum number of active pods required to handle workloads. Specifically, the peak pod count was reduced from 150 under the reactive HPA baseline to just 14 with InformerAutoScale, amounting to a 90.67% improvement. This substantial reduction underscores the framework's capability to manage resources conservatively without sacrificing service quality, thereby contributing to significant cost savings and lower energy consumption.

In terms of provisioning accuracy, the InformerAutoScale framework significantly outperformed reactive methods. Under-provisioning, which occurs when the system allocates fewer resources than needed, leading to resource starvation and SLA violations, was reduced from 23.4% to 5.2%, reflecting a 77.78% improvement. Over-provisioning, which results in unnecessary resource allocation and wasted computational capacity, decreased from 41.7% to 8.9%, representing a 78.66% improvement. These improvements highlight InformerAutoScale's ability to strike an effective balance between responsiveness and resource efficiency by more accurately adapting to workload fluctuations [155].

Most importantly, the framework significantly enhanced the Response Time SLA—a crucial indicator of user experience and system performance. The SLA adherence improved from 82% under the reactive HPA approach to 98% with InformerAutoScale, corresponding to a 19.51% increase. This improvement ensures that end users experience faster and more consistent response times, even when facing highly variable or bursty workloads, ultimately enhancing overall service reliability and customer satisfaction [156].

The core forecasting engine within InformerAutoScale also demonstrated strong predictive capabilities. The model achieved a Mean Absolute Error (MAE) of 12.3 requests per second and a Root Mean Squared Error (RMSE) of 18.7 requests per second. These error metrics validate the model's precision in estimating short-term workload demands, enabling timely and accurate scaling decisions. Furthermore, the forecasting engine attained an impressive 94.2% prediction accuracy for 5-minute forecast horizons, providing high confidence in proactive resource allocation. This precise forecasting is directly responsible for the observed improvements in SLA adherence and resource efficiency, illustrating the critical role of accurate demand prediction in modern autoscaling solutions [157].

10.6 Discussion and Insights

The *InformerAutoScale* framework demonstrates significant advantages of predictive resource allocation compared to traditional reactive autoscaling mechanisms in cloud-native environments. A particularly notable achievement is the **90.66% reduction** in the maximum number of pods required to handle workload peaks. This reduction translates directly into substantial cost savings and more efficient utilization of computational resources. By minimizing unnecessary resource provisioning, the framework effectively lowers operational expenses while also contributing to environmental sustainability through reduced energy consumption. From a performance perspective, proactive scaling is crucial in maintaining consistent and low response times, especially during sudden spikes in traffic. Unlike reactive systems, which adjust resources only after predefined thresholds are breached, *InformerAutoScale* anticipates future workload demands and adjusts resource allocation ahead of time. This predictive adjustment results in smoother system behavior, reduced latency, and improved adherence to service level agreements (SLAs).

In addition to predictive scaling, the incorporation of gradual scaling policies enhances overall system stability. Reactive autoscaling methods are often prone to rapid scale-in and scale-out cycles triggered by metric fluctuations—a phenomenon commonly referred to as *thrashing*. This behavior can lead to unnecessary pod churn and increased overhead on the orchestration system. The *InformerAutoScale* framework addresses this by implementing controlled, prediction-driven adjustments that limit abrupt resource changes within a single control cycle. Consequently, this approach reduces instability and lessens the operational burden on the Kubernetes control plane, fostering a more resilient and predictable autoscaling environment.

Despite these clear benefits, certain limitations remain that require consideration. One such challenge is the cold start latency associated with pod initialization, which typically ranges between 2 and 5 seconds. This delay can impact the responsiveness of time-sensitive applications, particularly during unexpected demand surges. Furthermore, while the forecasting model underpinning *InformerAutoScale* is robust for typical workload patterns, it may exhibit degraded performance when confronted with rare or anomalous *Black Swan* events that deviate significantly from historical trends. In such cases, fallback reactive mechanisms are necessary to maintain service continuity. Additionally, the framework requires periodic retraining of the forecasting model to adapt to evolving workload characteristics. Without timely updates, the system risks relying on outdated predictions, potentially resulting in suboptimal scaling decisions.

Looking ahead, there are several promising directions for future research to enhance the capabilities and applicability of the *InformerAutoScale* framework. One avenue involves developing hybrid scaling architectures that seamlessly combine proactive forecasting with reactive adjustments, thereby improving resilience and adaptability under diverse workload conditions. Another important extension is the integration of predictive autoscaling within *edge computing* environments, which introduces challenges such as network variability, resource heterogeneity, and geographically distributed management. Additionally, the application of *reinforcement learning* techniques could enable dynamic tuning of scaling policies, optimizing decisions based on real-time feedback and performance outcomes [158]. Finally, exploring predictive autoscaling strategies for *serverless (Function-as-a-Service)* platforms could yield significant efficiencies, particularly in mitigating cold start delays and improving invocation predictions in ephemeral computing models.

10.7 Key Takeaways

- **Introduction and Motivation:** The increasing complexity of cloud-native architectures demands more intelligent resource management solutions to handle dynamic workloads efficiently.
- **Challenges in Cloud-Native Architectures:** Traditional reactive autoscaling methods struggle with latency, resource over-provisioning, and under-provisioning, leading to inefficiencies and degraded user experience.
- **Reactive vs. Proactive Autoscaling:** Reactive autoscaling responds after thresholds are breached, often causing delayed reaction and instability, while proactive autoscaling leverages workload forecasting to anticipate demand, enabling smoother scaling decisions.
- **Autoscaling Models Overview:** Various predictive models such as LSTM and Transformer-based approaches provide forecasting capabilities, but often face trade-offs in accuracy and computational cost.

- **InformerAutoScale Framework:** A novel autoscaling solution integrating advanced forecasting using the Informer model with Kubernetes native scaling mechanisms to improve efficiency and responsiveness.
- **Architectural Overview:** The framework consists of data processing, forecasting, and execution layers, working cohesively to predict workload and enact controlled scaling actions.
- **Key Algorithms and Innovations:** InformerAutoScale employs state-of-the-art time series prediction and gradual scaling policies that minimize pod churn and ensure stability.
- **Methodology:** A robust data processing pipeline extracts and preprocesses metrics for forecasting, while a custom controller interfaces with Kubernetes HPA to implement scaling decisions.
- **Kubernetes Integration and Execution Layer:** The system enforces scaling actions through a Python-based controller with rate-limiting policies to prevent abrupt resource changes and maintain system stability.
- **Performance Evaluation:** Experiments on a realistic edge/cloud environment showed that InformerAutoScale significantly reduces maximum pod counts by over 90%, cuts under- and over-provisioning by more than 75%, and improves SLA adherence by nearly 20%.
- **Discussion and Insights:** Predictive scaling offers substantial benefits in cost reduction, resource efficiency, and response time consistency, while also highlighting areas such as cold start latency and model retraining as future challenges.
- **Future Directions:** Hybrid proactive-reactive architectures, edge computing integration, reinforcement learning enhancements, and serverless platform adaptation present promising opportunities to extend and improve autoscaling solutions.

10.8 Summary

This chapter presented the *InformerAutoScale* framework, a predictive autoscaling solution designed to overcome the limitations of traditional reactive methods in cloud-native environments. By leveraging advanced time-series forecasting with the Informer model, the framework anticipates workload demands, enabling more efficient and timely scaling decisions. The architecture integrates seamlessly with Kubernetes through a custom controller that enforces gradual scaling policies, ensuring system stability and reducing resource wastage. Comprehensive experiments demonstrated significant improvements in resource utilization, SLA adherence, and response times compared to baseline methods such as LSTM, Transformer models, and the native Kubernetes Horizontal Pod Autoscaler. Despite its advantages, challenges remain, including handling cold start latency and adapting to rare anomalous events, highlighting areas for future research. Overall, *InformerAutoScale* offers a promising step towards smarter, cost-effective, and resilient autoscaling in modern cloud-native systems.

10.9 Review Questions

1. What are the primary limitations of reactive autoscaling methods in cloud-native environments?
2. How does predictive autoscaling differ from reactive autoscaling, and what advantages does it offer?
3. Describe the role of the Informer model in the *InformerAutoScale* framework.
4. What are the key components of the data processing pipeline used in this autoscaling system?
5. How does Kubernetes integrate with the predictive autoscaling model in the framework?
6. Explain the purpose and function of the execution layer in the autoscaling architecture.
7. What performance metrics were used to evaluate the *InformerAutoScale* framework, and what improvements were observed?
8. Discuss the challenges related to cold start latency and anomalous workload patterns in predictive autoscaling.
9. How do gradual scaling policies help maintain system stability in the *InformerAutoScale* framework?
10. What future research directions are suggested to enhance predictive autoscaling frameworks?
11. Why is feature engineering important in the data processing pipeline, and what types of features were extracted?
12. How does time-series cross-validation improve model training for workload prediction?
13. Explain the significance of the custom metric `predicted_request_rate` in Kubernetes autoscaling.
14. What mechanisms are used to prevent overfitting during model training?
15. How does the InformerAutoScale framework balance between under-provisioning and over-provisioning?
16. In what ways does the framework contribute to environmental sustainability?
17. What are the roles of cAdvisor and Prometheus in the monitoring layer?
18. How does the InformerAutoScale system ensure responsiveness during sudden workload spikes?
19. What are the potential risks if the forecasting model is not periodically retrained?
20. Discuss how reinforcement learning could further improve autoscaling strategies.

Chapter 11
Installation and Setup

Abstract This chapter provides a practical guide for installing and setting up Kubernetes across different environments. It begins by outlining prerequisites for successful installation, followed by step-by-step procedures for deploying Kubernetes on CentOS 7 and Ubuntu systems. Minikube setup is covered as a lightweight option for local development and testing. Key aspects of network configuration and the Kubernetes dashboard setup are explained to ensure cluster accessibility and management. The process of adding worker nodes to an existing cluster is detailed, along with troubleshooting techniques for common installation issues. Best practices are highlighted to optimize reliability, security, and performance during setup. Finally, verification methods are presented to confirm that the installation and configuration are correctly implemented. This chapter equips readers with the practical knowledge needed to establish functional Kubernetes clusters in both development and production environments.

Keywords Kubernetes installation · Minikube · CentOS · Ubuntu · Network configuration · Dashboard · Cluster nodes · Troubleshooting

Installing Kubernetes requires thorough planning and precise configuration to create a stable, secure, and efficient cluster environment. Kubernetes is a powerful container orchestration tool, and its proper installation is essential for the smooth operation of containerized applications at scale [124]. This chapter will guide you through the process of setting up Kubernetes across various environments, such as CentOS, Ubuntu, and Minikube [159], which are common choices for different use cases. The chapter covers key aspects of the installation process, starting with the setup on CentOS and Ubuntu, two of the most widely used Linux distributions. It then explores Minikube, a tool that allows you to run Kubernetes clusters locally for testing and development purposes. The chapter also delves into network configuration, which is crucial for ensuring that the different components of the Kubernetes cluster communicate effectively [160]. Additionally, you will learn how to set up the Kubernetes dashboard, a web-based UI that helps you monitor and manage your Kubernetes cluster more easily. The chapter

B. Kumar et al., *Modern Kubernetes: From Core Concepts to Intelligent Autoscaling for Cloud Applications*, Studies in Autonomic, Data-driven and Industrial Computing, https://doi.org/10.1007/978-3-032-12972-7_11

also includes troubleshooting tips to help you address common installation issues, ensuring a smoother setup experience. Finally, it offers best practices to follow during installation, optimizing the performance, security, and scalability of your Kubernetes environment.

11.1 Prerequisites

Before starting with the installation of Kubernetes, it's essential to ensure that your system meets the necessary hardware, software, and network requirements [124]. Proper system configuration is crucial for a smooth and successful setup. Below are the prerequisites with detailed steps and commands to ensure that your system is properly prepared for Kubernetes deployment.

- Minimum System Requirements: These are the baseline requirements for setting up Kubernetes. While these are enough for a basic setup, production environments will need more resources to handle larger workloads effectively.
  - CPU: The system should have at least 2 virtual CPUs. However, for production environments, it's highly recommended to have 4 or more virtual CPUs to handle increased workload efficiently.
  - RAM: The system should have at least 2 GB of RAM, but for production environments, 8 GB or more is recommended to ensure Kubernetes and its workloads run smoothly.
  - Disk Space: We should have at least 20 GB of free storage. This ensures that there is enough room for Kubernetes, container images, and logs.
  - Operating System: The system should be running CentOS 7/8, Ubuntu 18.04/20.04, or a compatible Linux distribution. Additionally, if using Windows, WSL (Windows Subsystem for Linux) with PowerShell on these Linux distributions (such as CentOS) is supported.
  - Network: The system must have internet access for downloading necessary packages, and your firewall settings must be configured correctly to allow traffic for Kubernetes components.
- System Configuration: This section provides commands to configure the system and prepare it for Kubernetes installation.
  1. Disable Swap (Kubernetes does not work with swap enabled): Kubernetes doesn't work well if swap is enabled on the system, as it interferes with the way the kubelet works. Therefore, we need to disable swap permanently.

```
sudo swapoff -a
sudo sed -i '/ swap / s/^/#/' /etc/fstab
```

The first command sudo swapoff -a disables swap memory immediately. The second command comments out the swap entry in /etc/fstab, preventing it from being enabled after a reboot.

2. Enable IP Forwarding: IP forwarding allows packets to be forwarded between different network interfaces, which is required for communication between Kubernetes nodes.

```
sudo modprobe br_netfilter
echo 'net.ipv4.ip_forward = 1' | sudo tee -
    a /etc/sysctl.conf
sudo sysctl -p
```

The first command loads the br_netfilter module, which is necessary for Kubernetes' networking. The second command adds the necessary configuration to enable IPv4 packet forwarding in the sysctl.conf file. Finally, sudo sysctl -p applies the changes immediately.

3. Install Required Dependencies: Before installing Kubernetes, you need to install some basic dependencies like curl, apt-transport-https, and ca-certificates. These dependencies are required for downloading and verifying packages over HTTPS.

```
sudo apt update && sudo apt install -y curl apt-transport-
    https ca-certificates
```

The first command updates the package lists for the Ubuntu system. The second command installs the required packages to ensure that the system can securely download files via HTTPS and fetch Kubernetes binaries.

This section ensures that the system is adequately prepared to run Kubernetes. Disabling swap, enabling IP forwarding, and installing required dependencies are necessary steps before you can proceed with Kubernetes installation. Following these configurations will make sure that Kubernetes operates optimally on your system, without issues related to memory management or networking.

11.2 Installing Kubernetes on CentOS 7

To install Kubernetes on CentOS 7, there are several steps involved, starting with the installation of Docker, followed by Kubernetes package installation, and then initializing the master node [161]. Below are the detailed steps and necessary commands to install and configure Kubernetes successfully on a CentOS 7 machine.

- Step 1: Install Docker: Kubernetes relies on Docker as the container runtime to manage containers [128]. So, the first step is to install Docker on your CentOS 7 system.

```
sudo yum install -y yum-utils
sudo yum-config-manager --add-repo https://download.docker.
    com/linux/centos/docker-ce.repo
sudo yum install -y docker-ce docker-ce-cli containerd.io
sudo systemctl enable --now docker
```

- The first command installs yum-utils, which is required for managing repositories and configurations in CentOS.
- The second command adds Docker's official repository to your system, so you can install the latest version of Docker.
- The third command installs Docker CE (Community Edition), the Docker CLI, and containerd (which is used for managing containers).
- Finally, sudo systemctl enable –now docker ensures that Docker is set to automatically start on boot and starts it immediately.

- Step 2: Install Kubernetes Packages Now that Docker is installed, you need to install the necessary Kubernetes components: kubelet, kubeadm, and kubectl.

```
cat <<EOF | sudo tee /etc/yum.repos.d/kubernetes.repo
[kubernetes]
name=Kubernetes
baseurl=https://packages.cloud.google.com/yum/repos/
    kubernetes-el7
-x86_64
enabled=1
gpgcheck=1
repo_gpgcheck=1
gpgkey=https://packages.cloud.google.com/yum/doc/yum-key.gpg
    https://packages.cloud.google.com/yum/doc/rpm-package-key
    .gpg
EOF
sudo yum install -y kubelet kubeadm kubectl
sudo systemctl enable --now kubelet
```

The first command writes the Kubernetes repository configuration to the file. This repository, maintained by Google, hosts the Kubernetes packages specifically for CentOS, allowing the system to install and manage Kubernetes components easily. The second command installs the three main Kubernetes components: kubelet, kubeadm, and kubectl. The kubelet is the primary agent that runs on each node in the cluster and is responsible for managing containers and ensuring they are running as expected. The kubeadm tool is used to initialize the Kubernetes cluster and handle its configuration, while kubectl serves as the command-line tool used to interact with and manage the Kubernetes cluster, providing the user with a way to control and monitor the system. Finally, the systemctl command is used to ensure that the kubelet service is enabled to start automatically when the system boots. It also starts the service immediately, ensuring that Kubernetes can begin operating on the system without delay.

- Step 3: Initialize the Kubernetes Master Node: Once the necessary packages are installed, the next step is to initialize the Kubernetes master node using kubeadm. This will set up the control plane of the Kubernetes cluster.

```
sudo kubeadm init --pod-network-cidr
    =192.168.0.0/16
```

  The kubeadm init command initializes the Kubernetes master node. The –pod-network-cidr=192.168.0.0/16 flag specifies the CIDR block for pod networking (required for the network plugin you will install later). After running this command, Kubernetes will provide a join command that should be run on worker nodes to allow them to join the cluster. The output will look something like this:

```
kubeadm join <MASTER-IP>:<PORT> --token <TOKEN>
    --discovery-token-ca-cert-hash sha256:<HASH>
```

  Copy this command and save it for when you are ready to join worker nodes to the cluster.

These steps guide you through the installation of Kubernetes on a CentOS 7 system. First, Docker is installed, which is required by Kubernetes for container runtime. Then, the necessary Kubernetes components (kubelet, kubeadm, kubectl) are installed. Finally, the Kubernetes master node is initialized using kubeadm init, and the join command for worker nodes is generated. This sets up your control plane and prepares the system for adding worker nodes to the cluster.

11.3 Installing Kubernetes on Ubuntu

The process for installing Kubernetes on Ubuntu is similar to CentOS, but the commands and package management tools differ slightly. Ubuntu uses apt for package management [162]. This section will guide you through installing Docker, Kubernetes packages, and initializing the Kubernetes master node on an Ubuntu system.

- Step 1: Install Docker: Docker is essential for Kubernetes to run containers. To install Docker on Ubuntu, follow these steps:

```
sudo apt-get update
sudo apt-get install -y docker.io
sudo systemctl enable --now docker
```

  The first command updates the package lists for the system, ensuring you're installing the latest available versions of software. The second command installs Docker using the docker.io package available in Ubuntu's repository, and the third command ensures that Docker starts automatically on boot and starts it immediately without needing a restart.

- Step 2: Install Kubernetes Packages: Now that Docker is installed, we need to install Kubernetes components: kubelet, kubeadm, and kubectl. The following commands set up the Kubernetes repository and install the required packages.

```
sudo apt-get update
sudo apt-get install -y apt-transport-https curl
curl -s https://packages.cloud.google.com/apt/doc/apt-key.gpg
    | sudo apt-key add -
echo "deb http://apt.kubernetes.io/ kubernetes-xenial main" |
    sudo tee -a /etc/apt/sources.list.d/kubernetes.list
sudo apt-get update
sudo apt-get install -y kubelet kubeadm kubectl
sudo systemctl enable --now kubelet
```

  The first two commands update the system and install essential packages such as apt-transport-https and curl. These tools are required to access repositories over HTTPS, ensuring that the system can fetch packages from remote sources securely. The curl command is used to add the Google Cloud GPG key to the system. This key is important because it ensures that packages fetched from the Kubernetes repository are properly verified, maintaining the integrity and security of the software being installed. Next, the echo command is used to add the Kubernetes repository to Ubuntu's APT sources list. By doing this, the system is now configured to fetch Kubernetes components from this repository when performing package installations. Once the Kubernetes repository is added, the sudo apt-get update command is executed. This updates the package lists, ensuring that the system is aware of the newly added Kubernetes packages and can access them for installation.
- Step 3: Initialize the Kubernetes Master Node: Once the necessary Kubernetes packages are installed, the next step is to initialize the Kubernetes master node using kubeadm. This will set up the control plane for your cluster.

```
sudo kubeadm init --pod-network-cidr
    =192.168.0.0/16
```

  The kubeadm init command initializes the master node for your Kubernetes cluster. The –pod-network-cidr=192.168.0.0/16 flag specifies the CIDR block for pod networking. This is necessary to set up the networking for your pods, and this CIDR range can be customized based on the network plugin you plan to use. After executing this command, Kubernetes will display a join command that should be run on worker nodes to add them to the cluster. The output will look similar to:

```
kubeadm join <MASTER-IP>:<PORT> --token <TOKEN> --discovery-
    token-ca-cert-hash sha256:<HASH>
```

  Make sure to copy this join command and save it for when you are ready to add worker nodes to the cluster.

This section walks you through the installation of Kubernetes on Ubuntu. First, Docker is installed, which is required for container runtime. Then, the necessary Kubernetes components (kubelet, kubeadm, kubectl) are installed via APT, using Google's Kubernetes repository. Finally, the Kubernetes master node is initialized using kubeadm init, and the join command for worker nodes is generated. This prepares your system for setting up a Kubernetes cluster.

11.4 Minikube Setup

Minikube is a tool designed to create a single-node Kubernetes cluster that runs locally, which is ideal for development and testing purposes [159]. It simplifies the process of setting up a Kubernetes cluster without needing multiple machines or complex configurations. Below are the steps to install and set up Minikube on your local machine.

- Step 1: Install Minikube : To install Minikube, you need to download the latest Minikube binary and install it in a directory that's included in your system's $PATH. This step will show you how to do it on a Linux system.

```
curl -LO https://storage.googleapis.com/minikube/releases/
    latest/minikub
-linux-amd64
sudo install minikube-linux-amd64 /usr/local/bin/minikube
```

  The first command uses curl to download the latest Minikube binary from Google's storage and the second command installs the downloaded binary to /usr/local/bin/minikube, which is a common directory for executable files, ensuring Minikube is accessible globally from your command line.
- Step 2: Start Minikube: Once Minikube is installed, you can start the local Kubernetes cluster. Minikube will automatically set up a virtual machine (VM) and a Kubernetes control plane on your local machine.

```
          minikube start
```

  The minikube start command initiates the process of setting up a single-node Kubernetes cluster on your local machine. Minikube automatically configures all necessary components, including the Kubernetes control plane, and starts the VM to run the cluster. Minikube uses a lightweight VM or container (such as Docker, VirtualBox, etc.) to run the Kubernetes components.
- Step 3: Verify Installation: Once Minikube is running, you can verify the installation and check the status of your cluster by using kubectl, the Kubernetes command-line tool. It should be automatically configured to interact with the Minikube cluster.

```
kubectl cluster-info
kubectl get nodes
```

The kubectl cluster-info command shows the URL for accessing the cluster and provides basic information about the Kubernetes control plane. The kubectl get nodes command lists the nodes in your cluster. Since Minikube sets up a single-node cluster, it should show one node, typically named minikube.

Minikube is a simple and efficient way to run Kubernetes locally for development and testing. The setup involves installing Minikube, starting the Kubernetes cluster, and verifying that everything is working correctly with kubectl commands. By following these steps, you can quickly spin up a local Kubernetes cluster and start experimenting with Kubernetes features without the need for a full-scale deployment.

11.5 Network Configuration

Kubernetes relies on a Container Network Interface (CNI) plugin to handle networking within the cluster [39]. A CNI plugin is responsible for providing networking functionality to the pods, enabling them to communicate with each other. In this section, we'll set up Calico as the network plugin, which is one of the most popular CNI plugins due to its scalability, security, and flexibility [163]. Afterward, we'll verify the network setup to ensure everything is working as expected.

- Step 1: Installing Calico CNI Plugin: Calico is a high-performance networking and network security solution for containers. It's widely used in Kubernetes for pod networking.

```
kubectl apply -f https://docs.projectcalico.org/manifests/
    calico.yaml
```

- Step 2: Verifying Network Setup

```
kubectl get pods -n kube-system
```

  This command lists all the pods running in the kube-system namespace. It should show several pods, including the ones related to Calico, such as calico-node and calico-kube-controllers. If the pods are in the Running state, it means that the network is properly configured and Calico is functioning correctly.
- Alternative CNI Plugins: While Calico is a popular choice for Kubernetes networking, there are other CNI plugins that you can choose from, depending on your specific networking needs. These include:

- Flannel: A simple and easy-to-use CNI plugin, ideal for basic Kubernetes clusters with straightforward networking requirements.
- Cilium: A more advanced CNI plugin that uses eBPF (extended Berkeley Packet Filter) to provide high-performance networking with deep security features [164].
- Weave Net: A CNI plugin designed for simplicity and scalability. It is often used in environments where pod-to-pod communication is required across hosts without complex networking configurations.

Each of these plugins has its advantages, and the choice of which one to use depends on factors such as performance, security requirements, and the complexity of your Kubernetes network.

Network configuration in Kubernetes is crucial for ensuring seamless communication between pods. In this section, we installed the Calico CNI plugin to handle networking within the cluster. After installation, we verified the network setup by checking the status of the pods in the kube-system namespace. Alternative CNI plugins, such as Flannel, Cilium, and Weave Net, can also be used based on specific requirements for networking performance and security.

11.6 Kubernetes Dashboard Setup

The Kubernetes Dashboard is a web-based user interface that allows you to manage and monitor your Kubernetes cluster [165]. It provides a convenient way to view and interact with your cluster's resources, such as pods, services, deployments, and more. In this section, we'll walk through the process of deploying the dashboard, setting up access, and accessing it through a proxy.

- Step 1: Deploy the Dashboard: To deploy the Kubernetes Dashboard, you need to apply the recommended deployment manifest, which will create the necessary resources (such as deployments, services, etc.) to run the dashboard in your cluster.

```
kubectl apply -f https://raw.githubusercontent.com/kubernetes
    /dashboard/v2.3.1/
aio/deploy/recommended.yaml
```

  This command uses kubectl to apply the recommended manifest for the Kubernetes Dashboard. By running this command, Kubernetes resources needed for the dashboard are set up and configured automatically.
- The manifest defines several essential components, including:
  - Deployment: This is the pod that runs the dashboard application.
  - Service: It exposes the dashboard internally within the Kubernetes cluster, allowing other components to interact with it.

  – RBAC Resources: These define roles and permissions that control access to the dashboard, ensuring proper security and access management.

  Once the deployment is successful, the dashboard pod will be running in the kubernetes-dashboard namespace. This namespace helps isolate the dashboard resources and ensures they are properly managed within the cluster.
- Step 2: Start a Proxy to Access the Dashboard: To access the Kubernetes Dashboard, you need to start a kubectl proxy. This will create a secure connection between your local machine and the Kubernetes cluster, allowing you to access the dashboard in a web browser.

```
kubectl proxy
```

  The kubectl proxy command starts a local proxy to your Kubernetes cluster on your machine. By default, it listens on localhost:8001. The proxy ensures that your browser can securely access the Kubernetes API server and the dashboard, even though the dashboard service is running inside the cluster.
- Access the dashboard at: Once the proxy is running, you can access the dashboard through your web browser by visiting the following URL:

```
http://localhost:8001/api/v1/namespaces/
    kubernetes-
dashboard/services/https:kubernetes-dashboard:/
    proxy/
```

  This URL points to the Kubernetes Dashboard service running inside the cluster. The kubectl proxy ensures that you can access it locally through the localhost URL. When we navigate to this URL in your browser, you should see the Kubernetes Dashboard interface.
- Optional: Create an Admin User for Dashboard Access: By default, the Kubernetes Dashboard is not accessible to anyone without the proper access control permissions. To access the dashboard with full administrative privileges, you will need to create a service account with cluster-admin permissions.

```
kubectl create serviceaccount dashboard-admin-sa -n
    kubernetes-dashboard
kubectl create clusterrolebinding dashboard-admin-sa-binding
    --clusterrole=cluster-admin --serviceaccount=kubernetes-
    dashboard:dashboard-admin-sa
```

  The first command creates a service account (dashboard-admin-sa) in the kubernetes-dashboard namespace, and the second command binds the cluster-admin role to the service account, granting it full access to the cluster. After creating the service account, you can get the token for logging in:

```
kubectl -n kubernetes-dashboard create token dashboard-admin-
    sa
```

This command generates a login token that you can use to log into the Kubernetes Dashboard. Once you have the token, paste it into the login screen of the dashboard to access it with admin privileges.

The Kubernetes Dashboard is a powerful web interface for managing your cluster and its resources. This section guides you through deploying the Kubernetes Dashboard, starting a local proxy to access it, and providing the URL to log in. Additionally, we showed how to create an admin user with full cluster access for a seamless experience when interacting with the dashboard. By following these steps, you can easily monitor and manage your Kubernetes cluster from a browser-based UI.

11.7 Adding Worker Nodes to the Cluster

In a Kubernetes cluster, worker nodes run the containers and pods that make up your applications. After setting up the master node, you need to add worker nodes to the cluster to expand its capacity and functionality. Below are the steps to join a worker node to the Kubernetes cluster.

- Step 1: Get the Join Command from the Master Node: The master node generates a kubeadm join command, which is used to add worker nodes to the cluster. This command includes a unique token and the necessary information for the worker node to connect to the master node securely.

```
kubeadm token create --print-join-command
```

  This command is run on the master node. It generates the kubeadm join command, which contains: Token: A one-time token used for authenticating the worker node to the master and discovery Token CA Cert Hash: A security feature that ensures the worker node is joining the correct cluster. After running this command, the output will look something like this:

```
kubeadm join <master-ip>:6443 --token <token> --discovery-
    token-ca-cert-hash sha256:<hash>
```

  Copy this kubeadm join command, as it will be used on the worker node to join the cluster.
- Step 2: Join Worker Node: Next, you need to run the kubeadm join command on the worker node to add it to the cluster. This command establishes a connection to the master node and starts the process of registering the worker node in the cluster.

```
kubeadm join <master-ip>:6443 --token <token> --discovery-
    token-ca-cert-hash sha256:<hash>
```

master-ip: Replace this with the actual IP address of your master node. token: This is the token generated in Step 1. hash: This is the discovery token CA cert hash from Step 1, ensuring secure communication between the worker and master nodes. Once you execute this command on the worker node, the Kubernetes components will automatically configure the node and begin communicating with the master node.

- Verifying Worker Node Addition: After running the kubeadm join command on the worker node, you can verify the worker node has successfully joined the cluster by running the following command on the master node:

```
        kubectl get nodes
```

This command lists all the nodes in your cluster, including the master and any worker nodes. The worker node should show up in the list with the status Ready once it has successfully joined the cluster.

Adding worker nodes to a Kubernetes cluster involves running the kubeadm token create –print-join-command command on the master node to generate the join command, and then running the generated kubeadm join command on the worker node. Once the worker node joins the cluster, you can verify the successful addition by checking the node status with kubectl get nodes. This expands your cluster by adding more capacity and enabling the running of more containers and workloads.

11.8 Troubleshooting Kubernetes Installation

Even though Kubernetes is a powerful platform for managing containerized applications, issues can sometimes arise during installation or setup. In this section, we'll go over how to check the cluster status, inspect logs, and troubleshoot common issues during Kubernetes installation.

- Checking Cluster Status: The first step in troubleshooting any Kubernetes issue is to check the cluster status. By running a few kubectl commands, you can quickly gather insights into the health of the cluster.

```
            kubectl get nodes
            kubectl get pods --all-namespaces
```

This command displays the status of all nodes in the cluster, including the master and worker nodes. If a node is not in a Ready state, it indicates that the node might be experiencing an issue.

```
        kubectl get pods --all-namespace
```

This command lists all the pods running in the cluster across all namespaces. If any pods are not running or are in a CrashLoopBackOff or Pending state, it suggests there could be a problem with the pod's configuration or resources. By using these commands, you can quickly identify any nodes or pods that are not in a healthy state, which will help you narrow down the problem.

- Inspecting Logs: Logs are crucial for understanding what is happening inside the cluster and identifying errors. If a node, pod, or component isn't behaving as expected, checking the logs can give you more detailed information.

```
kubectl logs -f <pod-name> -n <namespace>
journalctl -xeu kubelet
```

This command follows the logs of a specific pod in a specific namespace. It is useful for identifying pod-related issues, such as crash loops or failing containers. journalctl -xeu kubelet: This command shows detailed logs for the kubelet service, which runs on every node and is responsible for managing pods and containers. If there are issues with the kubelet, such as failure to start or misconfiguration, the logs will show errors or warnings that can help diagnose the problem.

- Common Issues and Fixes: During installation or operation, you might encounter common issues that can prevent your cluster from functioning correctly. Below is a table that lists some typical issues, their causes, and suggested solutions:

```
Issue                    Cause                        Solution
Nodes in NotReady state Network misconfiguration Check CNI plugin
kubelet failing to start Wrong cgroup driver Match cgroup set
Pod stuck in Pending  Insufficient resources Check pod details
```

Nodes in NotReady state: This usually happens if there's a problem with networking or the CNI plugin. Make sure the CNI plugin (like Calico, Flannel, etc.) is properly configured and running on all nodes. Kubelet failing to start: One common cause is a mismatch in the cgroup driver used by Docker and Kubernetes. Ensure that both Docker and the kubelet are using the same cgroup driver (either cgroupfs or systemd). You can verify the cgroup driver in Docker with docker info and in kubelet using ps aux | grep kubelet. Pod stuck in Pending: This often occurs when the pod cannot find sufficient resources (CPU, memory) on the cluster nodes. You can describe the pod using kubectl describe pod <pod-name> to get detailed information about why it is pending. Also, check if any node is running out of resources or if there is any other configuration preventing scheduling.

Troubleshooting Kubernetes installation involves checking the cluster's health, inspecting logs for detailed errors, and identifying common issues that may arise during installation. By using commands like kubectl get nodes, kubectl logs, and journalctl, you can quickly diagnose problems with nodes, pods, or Kubernetes components. Additionally, understanding and fixing common issues like network misconfigurations, kubelet startup problems, and resource allocation issues will ensure a smoother Kubernetes experience.

11.9 Best Practices for Kubernetes Installation

When installing Kubernetes, it's important to follow certain best practices to ensure your cluster is secure, scalable, and maintainable. Below are some essential best practices to keep in mind during the installation and configuration process.

- Use Static IPs: Ensures nodes retain the same addresses: Assigning static IPs to your nodes helps prevent issues with dynamic IP address changes. Static IPs ensure that nodes consistently have the same network addresses, which is critical for stable communication between the master and worker nodes in the Kubernetes cluster.
- Enable Logs and Monitoring: Logs and monitoring are essential for maintaining the health and performance of a Kubernetes cluster. Setting up observability tools like Prometheus and Grafana enables you to track metrics, visualize system performance, and troubleshoot issues effectively. Prometheus collects and stores metrics, allowing you to monitor the health and performance of your cluster, while Grafana helps visualize these metrics through real-time dashboards [166]. This combination is crucial for identifying issues, analyzing trends, and keeping track of resource consumption. To implement this, you can install Prometheus to gather and store metrics from the Kubernetes API server, nodes, and pods. Then, integrate Grafana with Prometheus to visualize these metrics and create custom dashboards that will help you monitor the cluster's overall health and performance.
- Automate Deployment: Use Terraform or Ansible for cluster setup: Automating the deployment of your Kubernetes cluster ensures consistency, reduces the risk of human error, and simplifies the process of scaling and managing clusters across multiple environments [167]. Manual deployment is error-prone and time-consuming, but automation tools like Terraform and Ansible can help create, configure, and manage Kubernetes clusters in a reproducible way [168]. This allows you to quickly spin up new environments or scale your infrastructure efficiently. To implement this, you can use Terraform to define and provision the infrastructure required for your Kubernetes cluster, configuring cloud resources (such as EC2 instances) and Kubernetes settings using Infrastructure as Code (IaC). Additionally, Ansible can be used to automate the installation and configuration of Kubernetes components (like Docker, kubeadm, and kubectl), ensuring that the setup is consistent and repeatable. For example, a Terraform script can automate tasks like provisioning worker nodes, setting up the network, and installing Kubernetes, streamlining the entire deployment process.
- Set Up Role-Based Access Control (RBAC): Restrict permissions: RBAC (Role-Based Access Control) is a powerful tool in Kubernetes that allows you to manage permissions for users, groups, and service accounts [169]. By setting up RBAC, you ensure that users and services only have access to the resources they need, preventing unnecessary or unauthorized access. This is important because Kubernetes provides default access to cluster resources, but to adhere to the principle of least privilege, RBAC lets you restrict access to specific

namespaces, services, or operations. This enhances security by reducing the potential attack surface. To implement RBAC, you define Roles (or ClusterRoles) and RoleBindings (or ClusterRoleBindings) to assign specific permissions to users or service accounts. For example, you can create a read-only role that permits users to view resources but not modify them, or an admin role that grants full access to the entire cluster.

```
apiVersion: rbac.authorization.k8s.io/v1
kind: Role
metadata:
  namespace: default
  name: read-only
rules:
- apiGroups: [""]
  resources: ["pods"]
  verbs: ["get", "list"]
```

- Enable High Availability: Use multiple master nodes and load balancers: To ensure your Kubernetes cluster is fault-tolerant and can handle failures without downtime, you should deploy multiple master nodes and set up load balancing for high availability [38]. This is crucial because high availability ensures that your cluster remains operational even if one of the master nodes goes down. By deploying multiple master nodes, you eliminate a single point of failure, thus increasing the reliability of your cluster. Load balancers are essential for distributing traffic evenly among the master nodes, ensuring that the Kubernetes API server remains accessible at all times. To implement high availability, you can set up a load balancer in front of the master nodes to evenly distribute API requests. Additionally, configuring etcd clustering is important for data replication across multiple master nodes, which guarantees that the cluster's state remains consistent and available, even in the event of a node failure. For added resilience, it is recommended to deploy master nodes in different availability zones, offering geographical redundancy. A common example is using HAProxy or Nginx as load balancers to route traffic to multiple master nodes, helping ensure continuous API access and availability for your Kubernetes cluster.

Following best practices during Kubernetes installation ensures a secure, reliable, and maintainable cluster. By using static IPs, enabling logs and monitoring, automating deployments, setting up RBAC, and enabling high availability, you can ensure that your Kubernetes cluster is optimized for performance, scalability, and security. Implementing these best practices helps you maintain a robust Kubernetes environment, allowing for easier management, troubleshooting, and scaling [170].

11.10 Verification of Installation

Once you have completed the installation of your Kubernetes cluster, it is important to verify that everything is set up correctly and that the cluster is functioning as expected. This section covers basic cluster checks, running a sample deployment, and retrieving service details to confirm the installation.

- Basic Cluster Checks: To ensure the Kubernetes cluster is running properly, you should first perform some basic checks using kubectl commands. These checks will help you confirm that the cluster nodes and pods are healthy and that the Kubernetes API server is accessible.

```
kubectl get nodes
kubectl get pods --all-namespaces
kubectl cluster-info
```

  The First command lists all the nodes in your cluster and their statuses. Ensure that all nodes show up with a Ready status. If any node shows NotReady, there may be an issue with that node. The second command shows all the pods running in the cluster across all namespaces. Make sure that the essential system pods (like kube-system pods) are running and in the Running state and third command outputs the URL for the Kubernetes master API server and other critical information, confirming that the Kubernetes control plane is up and reachable. If these commands return expected results (e.g., all nodes are Ready, system pods are running), then the basic setup of your Kubernetes cluster is functional.
- Running a Sample Deployment: To further verify that your cluster is capable of running applications, you can create a simple deployment, such as an Nginx web server. This checks both the scheduling of pods and the accessibility of the deployed applications.

```
kubectl create deployment nginx --image=nginx
kubectl expose deployment nginx --port=80 --type=
    NodePort
```

  kubectl create deployment nginx –image=nginx: This command creates a deployment named nginx that runs an Nginx container. If the command succeeds, Kubernetes schedules the Nginx pod to run in the cluster. kubectl expose deployment nginx –port=80 –type=NodePort: This command exposes the Nginx deployment as a service, making it accessible on port 80. The –type=NodePort option means that the service will be exposed on a dynamic port on each worker node. By running this deployment, you're testing both the creation and management of pods as well as the ability of the cluster to expose services.
- Retrieve the service details using: After creating and exposing the deployment, you can retrieve the details of the service that Kubernetes created for the Nginx

deployment. This will show you the NodePort assigned to the service, which allows you to access the Nginx server from outside the cluster.

```
        kubectl get svc
```

This command lists all the services running in your cluster, including the Nginx service you just created. The output will display information such as the service name, type (NodePort), cluster IP, external IP (if applicable), and port numbers.

```
NAME   TYPE       CLUSTER-IP      EXTERNAL-IP   PORT(S)        AGE
nginx  NodePort   10.104.142.12   <none>        80:30001/TCP   2m
```

Here, the ‘PORT(S)‘ column shows that the Nginx service is exposed on port 30001 on all nodes. You can now access the Nginx server via any worker node’s IP address and the assigned NodePort.

To verify the successful installation of Kubernetes, you can begin by performing some basic cluster checks. Run commands like kubectl get nodes, kubectl get pods –all-namespaces, and kubectl cluster-info to ensure that the cluster and its components are healthy and functioning properly. These commands will provide information about the status of the nodes, the pods running across all namespaces, and general cluster health. Next, you can run a sample deployment to validate your installation further. Use kubectl create deployment to deploy an Nginx container, and then expose it with the kubectl expose command. This step confirms that the cluster is able to run applications and expose them externally, verifying that the basic deployment functionality is working. Finally, retrieve service details using kubectl get svc to confirm that the service was successfully created. Pay particular attention to the assigned NodePort, which will allow you to access the application from outside the cluster. By performing these verification steps, you can confirm that your Kubernetes installation is operational and ready to handle real workloads.

11.11 Key Takeaways

- A successful Kubernetes installation requires careful planning, considering hardware and software prerequisites to ensure cluster stability.
- Step-by-step installation guides for CentOS 7 and Ubuntu provide tailored approaches for setting up Kubernetes in different environments.
- Minikube serves as a lightweight alternative for local Kubernetes testing, allowing developers to experiment with cluster configurations.
- Proper network configuration is crucial for seamless communication between nodes and pods, ensuring reliability in deployments.

- The Kubernetes Dashboard provides a graphical interface for monitoring and managing cluster resources efficiently.
- Adding worker nodes is essential for scaling the cluster and distributing workloads effectively.
- Troubleshooting common installation issues, such as connectivity failures and misconfigurations, ensures a smooth setup process.
- Best practices, including security measures, resource allocation, and verification steps, help establish a robust and well-configured Kubernetes environment.

11.12 Summary

This chapter provided a detailed guide on installing and setting up Kubernetes across different environments, ensuring a smooth deployment process. It started by outlining the prerequisites, such as system requirements, package dependencies, and necessary configurations for a successful installation. Step-by-step instructions were given for installing Kubernetes on CentOS 7 and Ubuntu, highlighting differences in package managers and setup commands. For developers and testers, Minikube was introduced as an easy-to-use Kubernetes environment that allows running a single-node cluster locally. Network configuration was emphasized as a critical step to ensure proper communication between Kubernetes components, including setting up network plugins and firewall rules. Additionally, the chapter covered setting up the Kubernetes Dashboard, providing users with a visual management tool to monitor cluster activities, workloads, and performance. Adding worker nodes to a cluster was explored to enable scaling and workload distribution. Troubleshooting techniques were discussed to diagnose and resolve common installation issues, such as service failures, node connectivity problems, and incorrect configurations. Best practices were also covered, including securing cluster access, configuring role-based access control (RBAC), and verifying installation through diagnostic commands. By following the guidelines and best practices outlined in this chapter, users can successfully install, configure, and verify a Kubernetes environment that is secure, scalable, and optimized for production workloads.

11.13 Review Questions

- What are the minimum system requirements for installing Kubernetes?
- How do you install Kubernetes on CentOS 7 compared to Ubuntu?
- What is Minikube, and why is it useful for local Kubernetes development?
- What are the key components (kubeadm, kubelet, kubectl) needed for a Kubernetes setup?
- How do you configure a container runtime (Docker/containerd) for Kubernetes?

- Why is network configuration crucial, and which CNI plugins are commonly used?
- What steps are involved in setting up the Kubernetes Dashboard?
- How do you add and join worker nodes to a Kubernetes cluster?
- What are the most common installation errors, and how can they be fixed?
- How do you verify if Kubernetes is installed correctly?
- What security practices should be followed during Kubernetes installation?
- How can you deploy a test application to check cluster functionality?
- What logging/monitoring tools help troubleshoot Kubernetes setups?
- When should you use Minikube vs. a full cluster, and what are the trade-offs?
- What are the best practices for maintaining a stable Kubernetes installation?

Chapter 12
Kubernetes Deployment and Management

Abstract This chapter covers practical deployment and management techniques in Kubernetes. It begins with command-line interface (CLI) deployment methods, followed by YAML-based declarative deployments to illustrate flexible and reproducible configuration. Key operations such as scaling, verification, and troubleshooting are detailed, ensuring robust application management. Rolling updates are explained to highlight strategies for zero-downtime application upgrades, while deletion processes demonstrate safe resource cleanup. A step-by-step example of deploying Nginx provides a hands-on walkthrough for readers to understand the deployment workflow in practice. The chapter also addresses resource management techniques and the deployment of the Kubernetes dashboard for visual cluster monitoring and control. By combining theory with practical examples, this chapter equips readers with essential skills to deploy, manage, and maintain applications effectively in Kubernetes environments.

Keywords Deployment · CLI deployment · YAML · Scaling · Rolling updates · Troubleshooting · Nginx deployment

Kubernetes provides a robust framework for deploying and managing containerized applications, ensuring scalability, resilience, and efficient resource utilization [171]. This chapter explores various deployment strategies, including CLI-based and YAML-based deployments, highlighting the advantages of imperative and declarative approaches [172]. It covers essential aspects such as scaling applications to meet demand, verifying deployments, troubleshooting common issues, and managing rolling updates for seamless application upgrades. Additionally, it delves into resource management techniques to optimize CPU and memory allocation, as well as proper deletion practices to free up resources. A step-by-step walkthrough of deploying an Nginx application provides practical insights into Kubernetes workflows. Furthermore, the chapter discusses the Kubernetes dashboard, offering a visual interface for monitoring and managing deployments. By the end, readers will have a solid understanding of Kubernetes deployment and management, enabling them to efficiently orchestrate and maintain applications

B. Kumar et al., *Modern Kubernetes: From Core Concepts to Intelligent Autoscaling for Cloud Applications*, Studies in Autonomic, Data-driven and Industrial Computing, https://doi.org/10.1007/978-3-032-12972-7_12

in cloud-native environments [173]. This chapter covers Kubernetes deployment and management, focusing on CLI and YAML-based deployment methods. It explores scaling strategies, verification techniques, troubleshooting common issues, and performing rolling updates. Additionally, it provides step-by-step guidance on deploying an Nginx application in Kubernetes. The chapter also discusses resource management and dashboard deployment for monitoring and managing Kubernetes workloads effectively.

12.1 CLI Deployment

Kubernetes provides a powerful Command Line Interface (CLI) tool called kubectl, which enables users to interact with the cluster, manage applications, deploy workloads, expose services, and troubleshoot issues [174]. The CLI approach allows for quick and flexible deployment of applications without the need for complex configuration files.

Steps to Deploy Nginx Using CLI The following steps demonstrate how to deploy an Nginx application using kubectl:

- Create a deployment: A Kubernetes Deployment ensures that a specified number of application instances (pods) are running in the cluster. The kubectl create deployment command creates a deployment with the given name and container image

```
kubectl create deployment nginx --image=nginx
```

  kubectl create deployment command Creates a new deployment where nginx is the Name of the deployment and –image=nginx specifies the container image from Docker Hub.
- Expose the Deployment as a Service: Once the deployment is created, it must be exposed as a Service to allow external access. The kubectl expose command creates a service for the deployment.

```
kubectl expose deployment nginx --type=NodePort --
    port=80
```

  kubectl expose deployment nginx is the exposes the nginx deployment and –type=NodePort Makes the service accessible on a specific port and –port=80 → Exposes the application on port 80 inside the container.
- Access the Application: To access the running application, retrieve the allocated NodePort with the following command:

```
kubectl get svc nginx
```

This will output details about the service, including the NodePort, which can be used to access the application at:

```
http://<Node-IP>:<Node-Port>
```

where Node-IP is the IP address of any worker node in the cluster, and Node-Port is the dynamically assigned port.

Using the Kubernetes CLI, deploying an application is straightforward and efficient. The process involves creating a deployment, exposing it as a service, and retrieving the assigned NodePort to access the application. This method is ideal for quick testing and development environments where declarative YAML configurations are not necessary. However, for more complex deployments with defined configurations, YAML-based deployment is recommended.

12.2 YAML Deployment

YAML (Yet Another Markup Language) is widely used in Kubernetes for declarative configuration. It allows users to define resources in a structured and version-controlled manner [175]. Unlike CLI-based deployments, which are imperative, YAML configurations enable consistency and reusability, making them suitable for managing complex applications.

- Example YAML Deployment File (deployment.yaml) The following YAML file defines a Kubernetes Deployment for an Nginx application:

```yaml
apiVersion: apps/v1
kind: Deployment
metadata:
  name: nginx-deployment
spec:
  replicas: 3
  selector:
    matchLabels:
      app: nginx
  template:
    metadata:
      labels:
        app: nginx
    spec:
      containers:
      - name: nginx
        image: nginx
        ports:
        - containerPort: 80
```

Explanation of YAML Components:

- apiVersion: apps/v1 → Specifies the API version for the Deployment resource.
- kind: Deployment → Defines this resource as a Deployment.
- metadata.name: nginx-deployment → Assigns a name to the deployment.
- spec.replicas: 3 → Deploys three instances (pods) of the application.
- selector.matchLabels and template.metadata.labels → Ensure that the pods match the correct deployment for scaling and updates.
- spec.containers.image: nginx → Specifies the container image from Docker Hub.
- ports.containerPort: 80 → Exposes the application inside the container on port 80.

• Applying the YAML File: Once the YAML file is created, deploy it using the following command:

```
kubectl apply -f deployment.yaml
```

This command instructs Kubernetes to create or update the deployment based on the defined configuration.

Using YAML for Kubernetes deployments provides a structured and reusable approach to managing applications. It enables better version control, simplifies automation, and ensures consistency across environments. Unlike CLI-based deployments, YAML configurations allow teams to maintain infrastructure as code, improving scalability and operational efficiency.

12.3 Scaling

Scaling is a fundamental feature of Kubernetes that allows applications to handle varying levels of workload by increasing or decreasing the number of running pods [176]. Kubernetes supports horizontal scaling, meaning that additional instances of an application (pods) can be created or removed dynamically based on demand. Scaling improves resource efficiency, enhances fault tolerance, and ensures application availability during traffic spikes. Kubernetes provides two primary scaling methods: **Manual Scaling**—Users manually adjust the number of running pods and **Auto-Scaling**—Kubernetes automatically adjusts the number of pods based on CPU or custom metrics.

• Manual Scaling: Manual scaling allows users to explicitly define the number of pods in a deployment. This method is useful for controlled scaling during predictable workload changes or testing scenarios. Command to Scale Up or Down Manually:

```
kubectl scale deployment nginx --replicas=5
```

kubectl scale → Command to adjust the number of pods. deployment nginx → Specifies the deployment name (nginx in this case). –replicas=5 → Sets the number of running pods to five. After scaling, check the number of running pods using:

```
kubectl get pods
```

If five instances of the Nginx application are running, the scaling action was successful.

- Auto-Scaling (CPU-Based Scaling): Kubernetes offers Horizontal Pod Autoscaler (HPA), which automatically adjusts the number of pods based on CPU utilization. This method is ideal for handling fluctuating workloads efficiently [177]. Command to Enable Auto-Scaling:

```
kubectl autoscale deployment nginx --cpu-percent=50
    --min=2 --max=10
```

Where, kubectl autoscale → Enables automatic scaling for the specified deployment. deployment nginx → Targets the Nginx deployment. –cpu-percent=50 → Pods scale when CPU utilization exceeds 50% of allocated resources. –min=2 → Ensures a minimum of two pods are always running. –max=10 → Allows scaling up to a maximum of ten pods based on CPU demand.

- Verifying Auto-Scaling Configuration: After enabling auto-scaling, check the status using:

```
kubectl get hpa
```

This command provides details on the configured Horizontal Pod Autoscaler (HPA), including CPU utilization, target thresholds, and the current number of pods.

- Monitoring and Managing Scaling:
  - Check HPA Metrics:

```
kubectl describe hpa nginx
```

  Provides detailed information on the autoscaler, such as observed CPU usage and scaling activities.
  - Remove Auto-Scaling Configuration:

```
kubectl delete hpa nginx
```

Deletes the autoscaler and stops automatic scaling.

Kubernetes scaling ensures that applications can dynamically handle workload fluctuations. Manual scaling allows administrators to set the number of running pods explicitly, while auto-scaling adjusts the pod count based on CPU utilization, ensuring efficient resource usage. Auto-scaling helps prevent under-provisioning (which can cause performance issues) and over-provisioning (which leads to wasted resources). By leveraging Kubernetes's scaling capabilities, applications can maintain optimal performance under varying traffic loads.

12.4 Verification

Ensuring the successful deployment of an application in Kubernetes requires verification of running pods and services. Kubernetes provides a set of CLI commands to inspect the status of deployed resources, troubleshoot issues, and confirm that everything is functioning as expected. The key verification steps include checking running pods, reviewing service details, and inspecting specific pods in depth.

- Check Running Pods:

  ```
  kubectl get pods
  ```

  This command provides a list of all running pod instances in the cluster, displaying important details such as pod names, statuses (Running, Pending, CrashLoopBackOff, etc.), readiness status, restart counts, and uptime. If a pod is not running as expected, further debugging can be done using logs and descriptions.

  ```
  kubectl get svc
  ```

  Displays service details, including port mappings and IP addresses.
- Describe a Pod for More Information

  ```
  kubectl describe pod <pod-name>
  ```

Verification is a crucial step in Kubernetes deployment to ensure that applications are running correctly. By using commands like `kubectl get pods`, `kubectl get svc`, and `kubectl describe pod`, administrators can monitor the state of their resources, identify potential issues, and take corrective actions when necessary. These steps enhance the reliability and stability of applications deployed in a Kubernetes environment.

12.5 Troubleshooting

When an application is not functioning as expected in Kubernetes, various debugging tools help identify and resolve issues. Kubernetes provides commands to check logs, inspect deployment events, monitor pod status, and access containers for in-depth troubleshooting.

- Check Logs of a Pod

```
kubectl logs <pod-name>
```

  Retrieves logs from a specific pod to diagnose application-related issues such as errors, crashes, or unexpected behaviors.
- Check Events Related to a Deployment

```
kubectl describe deployment nginx
```

  Provides detailed information about a deployment, including configuration details, events, warnings, and errors, aiding in debugging failures or misconfigurations.
- Check Container Status in a Pod

```
kubectl get pods -o wide
```

  Lists all running pods with their IP addresses, node assignments, and statuses, helping to identify failed or pending pods.
- Exec into a Running Pod for Debugging

```
kubectl exec -it <pod-name> -- /bin/sh
```

  Opens an interactive shell inside a running container, allowing administrators to inspect logs, files, and processes for troubleshooting.

Kubernetes troubleshooting involves analyzing logs, inspecting deployment events, monitoring pod statuses, and accessing containers. These tools enable quick identification and resolution of issues, ensuring application stability and smooth cluster operation.

12.6 Rolling Updates

Rolling updates in Kubernetes enable zero-downtime deployments by incrementally replacing old application versions with new ones. This ensures continuous availability while reducing the risk of service disruptions [178].

- Edit the Deployment YAML: Modify the image field in deployment.yaml:

```
containers:
    - name: nginx
      image: nginx:1.21
```

  Modify the deployment configuration to specify a new image version. This change prepares Kubernetes to update the running application without downtime.
- Apply the Updated Deployment

```
kubectl apply -f deployment.yaml
```

  Applies the modified configuration, triggering Kubernetes to initiate a rolling update by gradually replacing old pods with new ones.
- Monitor the Update Process

```
kubectl rollout status deployment nginx
```

  Checks the status of the deployment update, ensuring that all pods transition successfully to the new version without issues.
- Rollback if Necessary

```
kubectl rollout undo deployment nginx
```

  Reverts the deployment to the previous stable version if the update encounters problems, ensuring minimal service disruption.

Rolling updates allow seamless application upgrades by gradually replacing old pods with new ones. Kubernetes ensures that updates progress smoothly, and if any issues arise, a rollback can quickly restore the previous version, maintaining system stability.

12.7 Deletion

In Kubernetes, resources such as deployments and services can be removed when they are no longer needed. Deleting these resources helps free up cluster resources and maintain a clean environment.

- To remove a deployment, use:

```
kubectl delete deployment nginx
```

  This command removes the specified deployment along with all running pods associated with it. This is useful when decommissioning an application or updating its configuration. This command deletes all running pods of the nginx deployment.

- To remove a service:

```
kubectl delete svc nginx
```

  Deletes the specified service, stopping traffic routing to the associated pods. This is necessary when removing or replacing exposed applications.

Kubernetes provides commands to delete deployments and services when they are no longer required. Removing a deployment stops its running pods, while deleting a service prevents access to the application. These actions help optimize cluster resource usage.

12.8 Example: Nginx Deployment (Step-by-Step Walkthrough)

This section provides a complete step-by-step guide to deploying, managing, and deleting an Nginx application in Kubernetes [179]. The following commands help create a deployment, expose it as a service, scale it, perform updates, and remove resources when no longer needed [170].

- Create Deployment

```
kubectl create deployment nginx --image=nginx
```

  Creates an Nginx deployment in Kubernetes, running an Nginx container.
- Expose the Deployment

```
kubectl expose deployment nginx --type=NodePort
    --port=80
```

  Exposes the deployment as a service, allowing external access on port 80.
- Check Running Pods and Services

```
kubectl get pods
kubectl get svc
```

  List all running pods and services to verify that the deployment and service are active.
- Scale Up the Deployment

```
kubectl scale deployment nginx --replicas=5
```

Increase the number of running Nginx pods to five for better availability.

- Perform a Rolling Update

```
kubectl set image deployment/nginx nginx=nginx
    :1.21
```

Updates the Nginx deployment to a newer image version without downtime.

- Rollback if Needed

```
kubectl rollout undo deployment nginx
```

Reverts the deployment to the previous version if issues arise.

- Delete the Deployment and Service

```
kubectl delete deployment nginx
kubectl delete svc nginx
```

Removes the Nginx deployment and service, cleaning up resources.

This example demonstrates the full lifecycle of an Nginx deployment in Kubernetes. It covers deployment creation, exposure, scaling, updating, rollback, and deletion, providing a practical understanding of Kubernetes resource management.

12.9 Resource Management

Kubernetes enables efficient resource allocation by setting CPU and memory limits for pods. This prevents resource contention and ensures fair distribution across workloads [180].

- Define Resource Limits in YAML

```
resources:
  requests:
    memory: "256Mi"
    cpu: "250m"
  limits:
    memory: "512Mi"
    cpu: "500m"
```

where requests → Specifies the minimum required CPU and memory resources for a pod. Kubernetes ensures the pod gets at least these resources. limits →Defines the maximum CPU and memory a pod can use, preventing it from consuming excessive resources.

- Apply the Updated Deployment

```
kubectl apply -f deployment.yaml
```

Applies the updated configuration, ensuring the resource constraints are enforced on the deployed application.

Kubernetes allows fine-grained control over CPU and memory usage through resource requests and limits. By defining these constraints in the YAML configuration, applications can efficiently utilize cluster resources while preventing overuse or starvation.

12.10 Dashboard Deployment

The Kubernetes Dashboard is a web-based user interface that allows users to manage and monitor cluster resources easily. It provides insights into deployments, services, pods, and other Kubernetes objects [181].

- Enable the Dashboard

```
kubectl apply -f https://raw.githubusercontent.com/kubernetes
    /dashboard/v2.0.0/aio/
deploy/recommended.yaml
```

Deploys the Kubernetes Dashboard, setting up the necessary resources in the cluster.
- Start the Dashboard Proxy

```
kubectl proxy
```

Starts a proxy server to securely access the Dashboard. The Dashboard can then be accessed via:

```
http://localhost:8001/api/v1/namespaces/kubernetes-dashboard/
    services/https:kubernetes-dashboard:/proxy/
```

- Get Access Token for Login

```
kubectl -n kube-system get secret | grep
    dashboard-token
kubectl -n kube-system describe secret <token-
    name>
```

Retrieves the authentication token required to log in to the Dashboard. Copy the token and use it to authenticate when accessing the UI.

The Kubernetes Dashboard provides a graphical interface for managing cluster resources. It can be deployed using a simple command, accessed through a proxy server, and secured using a token-based authentication system.

12.11 Key Takeaways

- Understanding various deployment strategies and best practices is essential for efficient application management in Kubernetes.
- Kubernetes supports deployment via the Command Line Interface (CLI) and declarative YAML configurations, enabling automation and repeatability.
- Scaling mechanisms, including manual and automated scaling, allow workloads to dynamically adjust to changing demands.
- Verification processes, such as checking pod statuses and logs, help ensure successful deployments and early issue detection.
- Rolling updates enable seamless application upgrades without downtime, maintaining service availability.
- Effective troubleshooting methods, such as inspecting events and logs, help resolve common deployment failures.
- Resource management using requests and limits optimizes CPU and memory utilization, ensuring efficient workload execution.
- A hands-on Nginx deployment example provided step-by-step guidance for setting up applications in Kubernetes.
- The Kubernetes Dashboard offers a visual interface for monitoring deployments and managing resources, streamlining operational workflows.
- Mastering these deployment techniques enhances scalability, reliability, and operational efficiency in Kubernetes environments.

12.12 Summary

This chapter provided a comprehensive guide to deploying and managing applications in Kubernetes, covering both CLI-based and YAML-based deployment methods. The chapter began by explaining the differences between imperative and declarative deployment approaches, highlighting the benefits of using YAML for consistency and automation. Scaling mechanisms, including horizontal and vertical scaling, were discussed to ensure that Kubernetes workloads can adapt to varying traffic and resource demands. The verification process was emphasized as a crucial step in deployment, detailing how to use commands such as kubectl get pods and kubectl logs to check deployment status and troubleshoot potential issues. Troubleshooting techniques, including checking pod events and error messages, were explored to address common deployment failures. Rolling updates were introduced as a strategy to minimize downtime during application upgrades, ensuring continuous service availability. A step-by-step example of deploying an Nginx application demonstrated practical deployment workflows, from defining deployment manifests to managing updates and scaling. The importance of resource management was also covered, explaining how to set CPU and memory limits to optimize performance and prevent resource exhaustion. Lastly, the Kubernetes Dashboard was introduced as a powerful visual tool for monitoring deployments, managing resources, and troubleshooting issues more efficiently. By following these best practices and leveraging Kubernetes' built-in tools, organizations can streamline their deployment processes, improve resource efficiency, and maintain high availability for containerized applications.

12.13 Review Questions

- What are the key differences between CLI Deployment and YAML Deployment in Kubernetes?
- How does the kubectl command facilitate Kubernetes Deployment and Management?
- What are the steps involved in deploying an application using YAML?
- How can you scale a deployment in Kubernetes, and what are the different scaling strategies?
- What methods can be used to verify if a deployment was successful?
- What common troubleshooting steps can be taken when a Kubernetes deployment fails?
- What is the purpose of rolling updates, and how do they ensure zero downtime during updates?
- How can you safely delete a deployment in Kubernetes?
- What are the step-by-step procedures for deploying Nginx on Kubernetes?
- How does Kubernetes handle resource management, and why is it important?
- What role does the Kubernetes Dashboard play in managing and monitoring clusters?
- How do you expose a deployment to external traffic using Services and Ingress?
- What are the differences between Horizontal Pod Autoscaler (HPA) and Vertical Pod Autoscaler (VPA) in scaling?
- How does Kubernetes handle configurations and secrets for deployments?
- What best practices should be followed when managing Kubernetes deployments for high availability and efficiency?

Chapter 13
Storage and Networking

Abstract This chapter provides a comprehensive overview of Kubernetes storage and networking concepts. It begins with storage fundamentals, including persistent volumes and claims, followed by networking basics that cover communication between pods and services. The role of Ingress is explained, along with network policies for controlling traffic flow and enhancing security. Practical examples guide readers through deploying an Ingress controller and creating a persistent volume claim for storage management. Troubleshooting techniques are provided to address common issues in storage and network configuration. Security considerations, such as securing Ingress with TLS, are discussed to ensure safe and reliable communication. Best practices are highlighted to optimize performance, reliability, and maintainability in Kubernetes clusters. By combining theory with hands-on examples, this chapter equips readers with the knowledge and practical skills needed to manage Kubernetes storage and networking effectively.

Keywords Kubernetes storage · Persistent volume · PVC · Networking · Ingress · Network policies · TLS · Ingress controller · Troubleshooting

Storage and networking formatted professionally with in-depth technical details. This version expands on Storage Overview, Types of Kubernetes Volumes, and their practical applications, benefits, and limitations. This chapter explores storage and networking in Kubernetes, covering fundamental concepts and practical implementations. It explains storage management through Persistent Volume Claims (PVCs) and troubleshooting techniques. The networking section introduces key concepts such as ingress, network policies, and securing ingress with TLS. A hands-on setup for deploying an ingress controller is provided, along with best practices to enhance security and efficiency in Kubernetes environments.

B. Kumar et al., *Modern Kubernetes: From Core Concepts to Intelligent Autoscaling for Cloud Applications*, Studies in Autonomic, Data-driven and Industrial Computing, https://doi.org/10.1007/978-3-032-12972-7_13

13.1 Storage Overview

Kubernetes provides robust storage mechanisms that allow applications to persist data beyond the lifespan of individual pods [182]. Unlike ephemeral storage, which is tied to the lifecycle of a pod and is lost when the pod is restarted or rescheduled, Kubernetes storage solutions ensure data continuity and availability. These storage options cater to diverse workloads, including stateful applications, distributed databases, and high-availability services [183].

Kubernetes Volumes In Kubernetes, a volume is an abstraction that allows containers within a pod to share storage [184]. Volumes are designed to persist data across container restarts within the same pod and, in some cases, across multiple pods. Kubernetes supports several volume types, each suited for specific use cases. Types of Kubernetes Volumes

- **emptyDir:** An emptyDir volume provides temporary storage that exists as long as the pod is running. It is created when a pod starts and deleted when the pod is removed from the node and **Use Case:** Ideal for caching, temporary file storage, or sharing files between containers in the same pod. **Limitations:** Data is lost when the pod is deleted or rescheduled. Example Usage:

```
apiVersion: v1
kind: Pod
metadata:
  name: example-emptydir
spec:
  containers:
  - name: app-container
    image: nginx
    volumeMounts:
    - mountPath: /cache
      name: cache-volume
  volumes:
  - name: cache-volume
    emptyDir: {}
```

- **hostPath:** The hostPath volume mounts a directory from the host node's filesystem into a pod. It is useful for accessing host resources directly and **Use Case:** IUsed in debugging, logging, or when the application needs access to specific host files. **Limitations:** Not recommended for production as it tightly couples the pod to a specific node, reducing portability. Example Usage:

```
apiVersion: v1
kind: Pod
metadata:
  name: example-hostpath
spec:
  containers:
```

```yaml
  - name: app-container
    image: nginx
    volumeMounts:
    - mountPath: /mnt/data
      name: host-volume
  volumes:
  - name: host-volume
    hostPath:
      path: /data
      type: Directory
```

- **Persistent Volume (PV) and Persistent Volume Claim (PVC):** Kubernetes introduces Persistent Volumes (PVs) and Persistent Volume Claims (PVCs) to manage long-term storage independently of pod lifecycles [184]. PVs abstract the underlying storage infrastructure, while PVCs request storage from available PVs. **Persistent Volume (PV):** Administrators define PVs, which represent physical storage. Supports various storage backends like NFS, AWS EBS, Azure Disks, and CephFS [185]. **Persistent Volume Claim (PVC):** Applications use PVCs to request storage without knowing the underlying storage provider. Example PV & PVC Configuration:

```yaml
apiVersion: v1
kind: PersistentVolume
metadata:
  name: example-pv
spec:
  capacity:
    storage: 5Gi
  accessModes:
    - ReadWriteOnce
  persistentVolumeReclaimPolicy: Retain
  hostPath:
    path: /mnt/data

apiVersion: v1
kind: PersistentVolumeClaim
metadata:
  name: example-pvc
spec:
  accessModes:
    - ReadWriteOnce
  resources:
    requests:
      storage: 5Gi

apiVersion: v1
kind: Pod
metadata:
  name: example-pvc-pod
spec:
  containers:
```

```
  - name: app-container
    image: nginx
    volumeMounts:
    - mountPath: /data
      name: storage-volume
  volumes:
  - name: storage-volume
    persistentVolumeClaim:
      claimName: example-pvc
```

- ConfigMaps & Secrets: Kubernetes provides ConfigMaps and Secrets to store configuration data separately from application code [89]. These are crucial for managing environment variables, sensitive information, and application settings. **ConfigMaps:** Stores non-sensitive configuration data such as application settings. Mounted as volumes or injected as environment variables. **Secrets:** Stores sensitive data such as passwords, API keys, and certificates. Encrypted at rest, offering a more secure alternative to ConfigMaps. Example Usage:

```
apiVersion: v1
kind: ConfigMap
metadata:
  name: example-config
data:
  database_url: "mongodb://mongo:27017"

apiVersion: v1
kind: Secret
metadata:
  name: example-secret
type: Opaque
data:
  password: cGFzc3dvcmQ=

apiVersion: v1
kind: Pod
metadata:
  name: example-config-pod
spec:
  containers:
  - name: app-container
    image: nginx
    env:
    - name: DATABASE_URL
      valueFrom:
        configMapKeyRef:
          name: example-config
          key: database_url
    volumeMounts:
    - name: secret-volume
      mountPath: /etc/secret
  volumes:
```

```
      - name: secret-volume
        secret:
          secretName: example-secret
```

Kubernetes provides multiple storage solutions to handle various application requirements. emptyDir and hostPath are useful for temporary storage and direct host access but have limitations in production. Persistent Volumes and Persistent Volume Claims enable stateful applications to manage data persistently across pod lifecycles [186]. ConfigMaps and Secrets help manage configuration and sensitive data securely. Understanding these storage options is essential for building resilient, scalable, and efficient Kubernetes applications.

13.2 Networking Basics

Kubernetes provides different service types to manage internal and external network communication within a cluster [39]. These services facilitate traffic routing, load balancing, and external accessibility for applications.

Kubernetes service types: Kubernetes provides four types of networking services:

- ClusterIP (Default)

```
kubectl expose deployment myapp --type=ClusterIP --port=80
```

  The default service type is accessible only within the cluster. Used for internal communication between pods.

- NodePort

```
kubectl expose deployment myapp --type=NodePort --port=80
kubectl get svc myapp  # Find the assigned NodePort
```

  Exposes the service on a static port (30000–32767) on each cluster node. Allows access from outside the cluster via NodeIP: NodePort.

- Load Balancer

```
kubectl expose deployment myapp --type=LoadBalancer --port=80
```

  Provision an external load balancer (if supported by the cloud provider). Automatically assigns an external IP for public access.

- ExternalName

```yaml
apiVersion: v1
kind: Service
metadata:
  name: my-external-service
spec:
  type: ExternalName
  externalName: example.com
```

Maps the service to an external domain name. Used for accessing external services without direct IP mapping.

Kubernetes networking provides multiple service types to facilitate communication between pods, nodes, and external systems. ClusterIP is used for internal communication, NodePort allows direct access to services, LoadBalancer integrates cloud-based load balancing, and ExternalName maps services to external domains. These options ensure seamless connectivity and traffic management within a Kubernetes cluster.

13.3 Ingress

Ingress is a Kubernetes API object that manages external HTTP(S) traffic and routes it to services within the cluster [187]. It provides a more efficient way to expose multiple services under a single domain using a set of routing rules.

- **Example: Basic Ingress Configuration**

```yaml
apiVersion: networking.k8s.io/v1
kind: Ingress
metadata:
  name: myapp-ingress
spec:
  rules:
  - host: myapp.example.com
    http:
      paths:
      - path: /
        pathType: Prefix
        backend:
          service:
            name: myapp
            port:
              number: 80
```

host: myapp.example.com → Routes traffic from this domain to the backend service. backend.service.name: myapp → Forwards requests to the “myapp”

service on port 80. pathType: Prefix → Matches traffic with the specified path prefix (/).
- Apply the Ingress Configuration

```
kubectl apply -f ingress.yaml
```

This command deploys the Ingress resource and enables external traffic routing based on the defined rules.

Ingress provides a structured way to manage HTTP(S) traffic into a Kubernetes cluster. It allows routing traffic to multiple services under a single domain, reducing the need for multiple LoadBalancer or NodePort services. Properly configured, Ingress simplifies external access and optimizes network management within Kubernetes.

13.4 Network Policies

Network policies in Kubernetes define rules that control how pods communicate with each other and with other network entities [188]. They enhance security by restricting unauthorized access within the cluster [160].

- Example: Allow Traffic Only from a Specific Namespace

```
apiVersion: networking.k8s.io/v1
kind: NetworkPolicy
metadata:
  name: allow-namespace
spec:
  podSelector:
    matchLabels:
      app: myapp
  ingress:
  - from:
    - namespaceSelector:
        matchLabels:
          name: trusted-namespace
```

podSelector.matchLabels.app: myapp → This policy applies to pods labeled myapp. trusted-namespace → Only allows traffic from pods within the trusted-namespace.
- Apply the Policy

```
kubectl apply -f network-policy.yaml
```

This command enforces the defined network restrictions within the cluster.

Network policies provide fine-grained control over pod communication, improving security by limiting unauthorized access. By defining rules, administrators can ensure that only trusted namespaces or specific pods can interact, reducing potential attack surfaces within the cluster.

13.5 Practical Setup: Deploying an Ingress Controller

An Ingress Controller is a key component in Kubernetes that manages external access to services within the cluster, typically handling HTTP and HTTPS traffic. It enables advanced routing capabilities, such as path-based routing, SSL termination, and load balancing.

- Install Nginx Ingress Controller : To deploy the Nginx Ingress Controller, apply the official Kubernetes manifest:

```
kubectl apply -f https://raw.githubusercontent.com/kubernetes
    /ingress-
        nginx/main/deploy/static/provider/cloud/deploy.yaml
```

  This command installs the required resources, including the Ingress controller deployment, service, and necessary configurations.
- Verify the Installation: After deploying the Ingress controller, check if the necessary pods are running in the ingress-nginx namespace:

```
        kubectl get pods -n ingress-nginx
```

The Ingress Controller plays a crucial role in Kubernetes by managing external traffic and routing it to the appropriate services within the cluster. It provides essential features such as load balancing, SSL termination, and name-based virtual hosting, but since Kubernetes does not include an Ingress Controller by default, it must be installed separately. Once deployed, users can define Ingress resources to control traffic flow efficiently. For instance, applying a simple Ingress rule using kubectl apply -f ingress.yaml enables structured and secure service access. Overall, setting up an Ingress Controller enhances traffic management and ensures seamless external connectivity for Kubernetes applications.

13.6 Storage Example: Persistent Volume Claim

Persistent storage in Kubernetes is essential for ensuring that data remains available even when pods restart or move across nodes [189]. This subsection covers the step-by-step process of creating and using Persistent Volumes and Persistent Volume Claims in Kubernetes.

- Create a Persistent Volume (PV): A Persistent Volume (PV) is a storage resource that exists independently of a pod's lifecycle. It can be provisioned manually by an administrator or dynamically based on storage classes. The following YAML definition creates a PV named pv-demo with a storage capacity of 5Gi and a ReadWriteOnce access mode. The hostPath specifies the location on the node where the data will be stored.

```
apiVersion: v1
kind: PersistentVolume
metadata:
  name: pv-demo
spec:
  capacity:
    storage: 5Gi
  accessModes:
    - ReadWriteOnce
  hostPath:
    path: "/mnt/data"
```

- Create a Persistent Volume Claim (PVC): A Persistent Volume Claim (PVC) allows a pod to request storage from available PVs. This ensures that applications receive the storage resources they need without manual intervention. The following YAML definition creates a PVC named pvc-demo that requests 2Gi of storage with a ReadWriteOnce access mode. Kubernetes will bind this PVC to a matching PV, if available.

```
apiVersion: v1
kind: PersistentVolumeClaim
metadata:
  name: pvc-demo
spec:
  accessModes:
    - ReadWriteOnce
  resources:
    requests:
      storage: 2Gi
```

- Apply the Configuration: Once the YAML files for the PV and PVC are created, apply them to the cluster using the following commands:

```
kubectl apply -f pv.yaml
kubectl apply -f pvc.yaml
```

These commands register the storage resources in the cluster. The PVC will automatically bind to an available PV that matches the specified storage requirements.

Persistent storage is crucial in Kubernetes for ensuring data persistence across pod restarts and rescheduling. This subsection covered the creation of a Persistent Volume (PV) and a Persistent Volume Claim (PVC), which allows applications to request and use storage efficiently. By applying the YAML configurations, Kubernetes dynamically manages storage allocation, providing a seamless experience for containerized applications.

13.7 Troubleshooting

Efficient troubleshooting in Kubernetes ensures smooth operation and quick resolution of issues related to networking, services, and pod connectivity. This subsection outlines key diagnostic commands to identify and resolve common problems.

- Check Service and Networking Issues: To investigate potential issues with a service, use the following command: This provides detailed information about the service, including its configuration, assigned cluster IP, ports, and potential error messages [190]. Useful for checking if the service is properly exposing the application.

```
kubectl describe svc myapp
```

- Debug Pod Connectivity Issues: If a pod is unable to communicate with a service, execute the following commands:

```
kubectl exec -it mypod -- /bin/sh
ping <service-name>
```

  The kubectl exec command allows access to a running pod for debugging purposes. The ping command helps verify if the service is reachable from within the pod.
- Check Network Policies: Network policies can restrict traffic between pods. To view the applied policies, use:

```
kubectl get networkpolicy
```

  This command lists all network policies applied within the cluster. It helps determine if a policy is unintentionally blocking communication between pods.

Troubleshooting in Kubernetes involves diagnosing networking issues, service misconfigurations, and pod connectivity problems. Using kubectl describe svc, kubectl exec, and kubectl get networkpolicy, administrators can identify and resolve issues efficiently. These commands provide valuable insights into service status, pod connectivity, and network policy enforcement, ensuring smooth cluster operation.

13.8 Security: Securing Ingress with TLS

Securing Kubernetes Ingress with TLS (Transport Layer Security) enhances security by encrypting HTTP(S) traffic between users and services [191]. This section covers TLS certificate generation, configuring Ingress for HTTPS, and applying the secure configuration.

- Generate TLS Certificates: To create self-signed TLS certificates for securing Ingress, use:

```
openssl req -x509 -nodes -days 365 -newkey rsa:2048 -keyout tls.key -out tls.crt -subj "/CN=myapp.example.com"
kubectl create secret tls my-tls-secret --cert=tls.crt --key=tls.key
```

  The openssl command generates a TLS certificate (tls.crt) and a private key (tls.key). The kubectl create secret command stores the TLS certificate as a Kubernetes secret named my-tls-secret.
- Secure Ingress with TLS : A secure Ingress configuration is defined in a YAML file (secure-ingress.yaml):

```
apiVersion: networking.k8s.io/v1
kind: Ingress
metadata:
  name: secure-ingress
spec:
  tls:
  - hosts:
    - myapp.example.com
    secretName: my-tls-secret
  rules:
  - host: myapp.example.com
    http:
      paths:
      - path: /
        pathType: Prefix
        backend:
          service:
            name: myapp
            port:
              number: 443
```

  The tls section binds the TLS certificate (my-tls-secret) to the specified hostname (myapp.example.com). The Ingress routes HTTPS requests to the myapp service on port 443.
- Apply the Secure Ingress : To enable TLS encryption, apply the Ingress YAML configuration:

```
kubectl apply -f secure-ingress.yaml
```

  This command registers the secure Ingress configuration with Kubernetes.

Securing Ingress with TLS ensures encrypted communication between external users and Kubernetes services. By generating TLS certificates, creating a Kubernetes secret, and configuring a secure Ingress resource, administrators can enable HTTPS traffic. This improves security, protecting sensitive data and preventing unauthorized access.

13.9 Best Practices

Implementing best practices in Kubernetes ensures a secure, scalable, and well-organized cluster environment. This section outlines key strategies, including namespace isolation, role-based access control (RBAC), and pod security policies [192].

- Use Namespaces for Isolation: Namespaces help isolate different environments (e.g., development and production) to improve security and resource management.

```
kubectl create namespace dev
kubectl create namespace prod
```

  The dev namespace is used for development workloads. The prod namespace is for production workloads.
- Enable RBAC (Role-Based Access Control): RBAC restricts user and application access based on roles, preventing unauthorized actions.

```
apiVersion: rbac.authorization.k8s.io/v1
kind: Role
metadata:
  name: pod-reader
  namespace: dev
rules:
- apiGroups: [""]
  resources: ["pods"]
  verbs: ["get", "list"]
```

  The pod-reader role allows only read access (get and list) to Pods in the dev namespace. This ensures that users have limited access to resources based on their roles.
- Implement Pod Security Policies: Pod Security Policies define security requirements for running containers, such as restricting privileged access.

```
kubectl apply -f pod-security.yaml
```

  This applies a security policy (pod-security.yaml) that can enforce security constraints on Kubernetes Pods.

Following best practices in Kubernetes enhances security, resource isolation, and access control. Using namespaces organizes workloads, RBAC limits permissions based on user roles, and pod security policies enforce security constraints. Implementing these strategies helps maintain a secure, well-structured, and efficient Kubernetes environment.

13.10 Key Takeaways

- Persistent Volumes (PV) and Persistent Volume Claims (PVC) enable stateful applications to retain data across pod restarts, ensuring reliable storage management.
- Networking fundamentals, including ClusterIP, NodePort, and LoadBalancer services, facilitate seamless communication between pods and external systems.
- The Ingress resource optimizes external traffic management by defining routing rules and enabling efficient load balancing.
- Network Policies help control traffic flow between pods, enhancing security and isolation within the cluster.
- A practical setup using Ingress Controllers demonstrated dynamic traffic management for Kubernetes applications.
- Troubleshooting techniques for storage and network issues, such as misconfigured Ingress rules or unbound Persistent Volume Claims, improve system reliability.
- Security best practices, including securing Ingress with TLS, enforcing Role-Based Access Control (RBAC), and implementing Pod Security Policies, help protect Kubernetes environments.
- Implementing these storage and networking strategies enhances application performance, security, and scalability in Kubernetes deployments.

13.11 Summary

This chapter provided an in-depth look at Kubernetes storage and networking, outlining essential components, configurations, and best practices. It began with an overview of storage management, focusing on Persistent Volumes (PV) and Persistent Volume Claims (PVC), which enable stateful workloads to persist data independently of pod lifecycles. The discussion then shifted to networking fundamentals, covering different Kubernetes service types—ClusterIP for internal communication, NodePort for external access, and LoadBalancer for cloud-based routing. The chapter also introduced Ingress resources, explaining their role in managing external HTTP/S traffic and optimizing request routing with rules and load balancing. A practical setup of deploying an Ingress Controller was included, demonstrating how to configure traffic flows dynamically. Security considerations, such as securing Ingress with TLS encryption, implementing Role-Based Access Control (RBAC), and enforcing Pod Security Policies, were highlighted as essential measures to protect Kubernetes workloads. Network Policies were explored in detail, emphasizing their importance in restricting unauthorized communication between pods and ensuring application security. Troubleshooting techniques were provided to address common networking and storage issues, including debugging unbound PVCs and misconfigured Ingress rules. By following these guidelines,

Kubernetes administrators can ensure robust storage solutions, efficient network communication, and a secure infrastructure, enabling scalable and high-performing cloud-native applications.

13.12 Review Questions

- What are the key differences between CLI deployment and YAML deployment in Kubernetes?
- How do you create a basic deployment using kubectl commands?
- What essential components must be included in a YAML deployment file?
- How does Kubernetes implement automatic scaling of deployments?
- What commands are used to manually scale deployments up or down?
- How do you verify that a deployment was successfully created and is running properly?
- What are the most common troubleshooting steps for failed deployments?
- How does Kubernetes handle rolling updates to minimize downtime?
- What command initiates a rollback to a previous deployment version?
- How do you safely delete deployments and their associated resources?
- Walk through the complete process of deploying an Nginx web server in Kubernetes.
- How do you manage resource allocation (CPU/memory) for deployments?
- What are resource requests and limits, and why are they important?
- How do you deploy and access the Kubernetes Dashboard?
- What are the key benefits of using the Dashboard versus CLI for deployment management?

Chapter 14
Security and Monitoring

Abstract This chapter addresses security and monitoring practices in Kubernetes to ensure safe and reliable cluster operations. It begins with an overview of Kubernetes security principles, followed by authentication and authorization mechanisms that control access to cluster resources. Monitoring strategies are discussed, including dashboard monitoring, logging, and the use of specialized tools to track cluster health and performance. Best practices are highlighted to strengthen security and improve observability across deployments. A real-world case study, the Tesla attack, illustrates potential threats and mitigation strategies. The chapter also introduces security tools and explores CI/CD security measures to protect deployment pipelines. By combining theoretical concepts with practical examples, this chapter equips readers with essential knowledge to secure Kubernetes environments effectively while maintaining visibility and operational insight for proactive management.

Keywords Kubernetes security · Monitoring · Authentication · Authorization · Logging · CI/CD · Security tools

Kubernetes security and monitoring are crucial for maintaining a stable, secure, and well-functioning cluster. Security focuses on protecting cluster resources, applications, and data from unauthorized access, vulnerabilities, and runtime threats. Monitoring helps detect issues, track system performance, and ensure the cluster remains in optimal health. This chapter focuses on security and monitoring in Kubernetes, detailing authentication, authorization, and security best practices. It explores monitoring techniques, including dashboard monitoring, logging, and essential tools for observability. A case study on the Tesla attack highlights real-world security challenges. Additionally, it covers CI/CD security and the role of security tools in safeguarding Kubernetes environments.

B. Kumar et al., *Modern Kubernetes: From Core Concepts to Intelligent Autoscaling for Cloud Applications*, Studies in Autonomic, Data-driven and Industrial Computing, https://doi.org/10.1007/978-3-032-12972-7_14

14.1 Security Overview

Security in Kubernetes is essential to safeguard the cluster from unauthorized access, vulnerabilities, and runtime threats. As Kubernetes is widely used for container orchestration, ensuring robust security measures is crucial to maintaining system integrity and data protection. Several key aspects contribute to Kubernetes security, including access control, image security, and runtime security [193, 194]. Access control plays a significant role in restricting unauthorized access to Kubernetes resources. One of the primary mechanisms for managing access is Role-Based Access Control (RBAC), which defines permissions for users and services. RBAC allows administrators to specify who can perform specific actions on resources within the cluster, ensuring that only authorized entities have the necessary privileges [192]. This minimizes security risks by preventing unintended modifications or unauthorized operations. Image security is another critical component of Kubernetes security. Before deploying containers, it is essential to scan container images for vulnerabilities. Using tools such as container vulnerability scanners, organizations can detect and mitigate security issues early in the development cycle. Ensuring that only trusted images are deployed reduces the risk of running containers with known security flaws, thereby strengthening the overall security posture of the cluster [195]. Runtime security focuses on protecting applications while they are running within Kubernetes. This involves monitoring application behavior, detecting anomalies, and enforcing security policies to prevent malicious activities. Tools and frameworks such as Kubernetes security policies, intrusion detection systems, and runtime monitoring solutions help ensure that applications operate securely. By continuously observing workloads and responding to threats in real time, runtime security enhances the resilience of Kubernetes environments against potential attacks.

- Role-Based Access Control (RBAC): RBAC helps restrict access to cluster resources based on user roles and permissions. It prevents unauthorized users from making changes or accessing sensitive data. Example: Creating a Role and RoleBinding: A Role in Kubernetes defines permissions for resources within a specific namespace. The following YAML file creates a Role that allows users to view pod information inside the dev namespace.

```yaml
apiVersion: rbac.authorization.k8s.io/v1
kind: Role
metadata:
  namespace: dev
  name: pod-reader
rules:
- apiGroups: [""]
  resources: ["pods"]
  verbs: ["get", "list"]
```

  This Role allows users to get and list pods within the dev namespace. It does not grant create, delete, or update permissions, ensuring read-only access. Next, the RoleBinding links the Role to a user, allowing access based on predefined permissions. The following YAML file binds the pod-reader Role to the user "john" in the dev namespace:

```
apiVersion: rbac.authorization.k8s.io/v1
kind: RoleBinding
metadata:
  name: pod-reader-binding
  namespace: dev
subjects:
- kind: User
  name: "john"
  apiGroup: rbac.authorization.k8s.io
roleRef:
  kind: Role
  name: pod-reader
  apiGroup: rbac.authorization.k8s.io
```

The Role Binding ensures that the user "john" can only view pods within the dev namespace. The roleRef links the pod-reader Role to the binding, ensuring limited permissions.

- Apply Role and Binding: Once the Role and RoleBinding configurations are created in YAML files (role.yaml and rolebinding.yaml), apply them using the following commands:

```
        kubectl apply -f role.yaml
        kubectl apply -f rolebinding.yaml
```

- Image Security: Container images should be scanned for vulnerabilities before deployment. Attackers can exploit known vulnerabilities if images are not securely managed. **Scan Images with Trivy:** Trivy is a security scanner for container images that detects vulnerabilities in dependencies [196].

```
        trivy image myapp:latest
```

Displays a list of vulnerabilities found in the container image. Categorizes them based on severity: Critical, High, Medium, and Low. Provides recommendations for fixing vulnerabilities.

- Securing Ingress with TLS: TLS (Transport Layer Security) encrypts traffic to prevent data interception.
  - Generating TLS Certificates

```
openssl req -x509 -nodes -days 365 -newkey rsa:2048 -keyout
     tls.key -out tls.crt -subj "/CN=myapp.example.com"
kubectl create secret tls my-tls-secret --cert=tls.crt --
    key=tls.key
```

– Configuring Secure Ingress

```yaml
apiVersion: networking.k8s.io/v1
kind: Ingress
metadata:
  name: secure-ingress
spec:
  tls:
  - hosts:
    - myapp.example.com
    secretName: my-tls-secret
  rules:
  - host: myapp.example.com
    http:
      paths:
      - path: /
        pathType: Prefix
        backend:
          service:
            name: myapp
            port:
              number: 443
```

– Applying the Secure Ingress

```
kubectl apply -f secure-ingress.yaml
```

Ensuring security in a Kubernetes cluster is crucial to protect against unauthorized access and potential vulnerabilities. Role-Based Access Control (RBAC) plays a key role in managing user and service permissions, ensuring that only authorized entities can access specific resources. Image security is another critical aspect, requiring container images to be scanned for vulnerabilities before deployment to prevent running compromised workloads. Additionally, securing ingress with TLS helps encrypt data in transit, protecting communication between external users and services within the cluster. Implementing these security measures strengthens the overall security posture of the Kubernetes environment.

14.2 Monitoring in Kubernetes

Monitoring is essential to detect performance issues, track resource usage, and troubleshoot failures. Kubernetes provides several monitoring tools and techniques.

- Monitoring Cluster Resources: To check resource usage (CPU, memory, and storage), use the following commands:

```
kubectl top nodes
kubectl top pods
```

- Logging with Kubectl: Logs help diagnose issues in Kubernetes workloads. Use the command:

```
kubectl logs my-pod
```

- Using Prometheus and Grafana: Prometheus collects metrics from cluster components [197]. Deploy Prometheus in Kubernetes using:

```
kubectl apply -f prometheus-deployment.yaml
```

  - Deploy Prometheus in Kubernetes:

    ```
    kubectl apply -f prometheus-deployment.
        yaml
    ```

  - Access Prometheus Web UI

    ```
    kubectl port-forward svc/prometheus
        9090:9090
    ```

    Open http://localhost:9090 in a browser.
  - Deploy Grafana in Kubernetes

    ```
    kubectl apply -f grafana-deployment.
        yaml
    ```

  - Access Grafana Web UI

    ```
    kubectl port-forward svc/grafana
        3000:3000
    ```

    Open http://localhost:3000 in a browser.

Effective monitoring is essential for maintaining the performance and reliability of a Kubernetes cluster. Using kubectl commands, administrators can check resource utilization at both the node and pod levels, gaining real-time insights into CPU, memory, and storage consumption. Prometheus serves as a powerful metrics collection tool, gathering performance data from various cluster components to facilitate proactive issue detection. Complementing this, Grafana provides visualization capabilities, allowing users to analyze metrics through interactive dashboards [198]. Together, these monitoring tools enable better observability, ensuring the smooth operation of the Kubernetes cluster.

14.3 Authentication

Kubernetes authentication ensures that only authorized users and services can access cluster resources [199]. Authentication mechanisms are essential to securing Kubernetes clusters by verifying identities before granting access.

- Common Authentication Methods:
  - Service Account Tokens—Kubernetes automatically creates service accounts for pods, assigning them unique tokens to authenticate API requests.
  - Client Certificates—Kubernetes uses TLS certificates to secure authentication. Administrators generate client certificates and configure Kubernetes API servers to recognize them.
  - OpenID Connect (OIDC)—OIDC allows integration with external identity providers like Google, GitHub, and Keycloak. This is useful for organizations using OAuth-based authentication.
- Example: OIDC Authentication via Google—To integrate Kubernetes authentication with Google's OAuth service, create an OIDC configuration using a ConfigMap.

```
apiVersion: v1
kind: ConfigMap
metadata:
  name: oidc-config
data:
  issuer: "https://accounts.google.com"
  clientID: "your-client-id"
  clientSecret: "your-client-secret"
```

  This configuration specifies Google as the identity provider and includes the client ID and secret for authentication [200].
- Apply Configuration: Apply Configuration Using kubectl Run the following command to apply the OIDC configuration to the Kubernetes cluster:

```
        kubectl apply -f oidc-config.yaml
```

  This ensures that the cluster is configured to authenticate users via Google OAuth.

Kubernetes authentication mechanisms secure cluster access through various methods, including service accounts, client certificates, and OpenID Connect. OIDC integration allows organizations to leverage external identity providers for seamless authentication.

14.4 Authorization

Authorization determines which users or services can perform specific actions on Kubernetes objects. After authentication, Kubernetes checks permissions using an authorization mechanism to enforce security policies.

- RBAC Example: RBAC is a common authorization method in Kubernetes, allowing administrators to define fine-grained permissions for users and services. Example: Grant User Access to Pods Using RBAC To grant a user the ability to view and list pods, define a ClusterRole with appropriate permissions.

```
apiVersion: rbac.authorization.k8s.io/v1
kind: ClusterRole
metadata:
  name: pod-access
rules:
- apiGroups: [""]
  resources: ["pods"]
  verbs: ["get", "list", "watch"]
```

  This ClusterRole allows users to retrieve (get), list (list), and monitor (watch) pod resources.
- Apply the RBAC Role Use kubectl to apply the role configuration:

```
kubectl apply -f role.yaml
```

  This registers the role in the cluster, enabling assigned users to interact with pods.

Kubernetes authorization enforces access control policies using RBAC, ensuring users and services operate within defined permissions. Proper authorization prevents unauthorized access and secures Kubernetes resources.

14.5 Monitoring Tools in Kubernetes

Monitoring is essential in Kubernetes to track cluster health, resource utilization, and performance metrics. Prometheus and Grafana are widely used monitoring tools for Kubernetes.

- Prometheus Metrics Collection: Prometheus is an open-source monitoring system that collects and stores time-series data, enabling real-time metric analysis [126].

– Install Prometheus: Apply the Prometheus Operator bundle to deploy Prometheus in the cluster:

```
kubectl apply -f https://raw.githubusercontent.com/
    prometheus-operator/prometheus
operator/main/bundle.yaml
```

This deploys Prometheus components, including the Prometheus server, Alertmanager, and exporters.

– Query Metrics: Expose the Prometheus UI locally using port forwarding:

```
kubectl port-forward svc/prometheus-
    service 9090:9090
```

Access the Prometheus web interface at http://localhost:9090 to query metrics.

• Grafana: Grafana is a popular visualization tool that integrates with Prometheus to create real-time dashboards [201].

– Install Grafana Deploy Grafana in Kubernetes:

```
kubectl apply -f https://raw.githubusercontent.com/grafana/
    grafana/main/
deploy/kubernetes.yaml
```

This sets up Grafana with a default configuration.

– Access Grafana Dashboard: Forward the Grafana service port to access the UI:

```
kubectl port-forward svc/grafana
    3000:3000
```

Visit http://localhost:3000 to explore dashboards and visualize metrics.

Kubernetes monitoring tools like Prometheus and Grafana provide powerful insights into cluster performance. Prometheus collects and queries metrics, while Grafana enables real-time visualization, enhancing observability and troubleshooting.

14.6 Dashboard Monitoring

The Kubernetes Dashboard provides a web-based user interface for managing and monitoring cluster resources. It allows users to view workloads, deployments, and resource utilization in real time.

- Install Kubernetes Dashboard: Deploy the Kubernetes Dashboard using the official YAML manifest:

```
kubectl apply -f https://raw.githubusercontent.com/kubernetes
    /dashboard/v2.3.1/aio/
deploy/recommended.yaml
```

  This installs the necessary components, including the dashboard UI, backend services, and role-based access control (RBAC) configurations.
- Access the Kubernetes Dashboard: To access the dashboard, start a proxy server:

```
        kubectl proxy
```

  Then, navigate to the following URL in a browser:

```
http://localhost:8001/api/v1/namespaces/kubernetes-dashboard/
    services/
https:kubernetes-dashboard:/proxy/
```

  This opens the dashboard, allowing you to monitor cluster activity, inspect workloads, and manage resources interactively.

The Kubernetes Dashboard provides a user-friendly interface for managing and monitoring clusters. It simplifies resource visualization, making it easier to track performance, troubleshoot issues, and perform administrative tasks.

14.7 Logging

Logging is essential for monitoring, debugging, and analyzing Kubernetes workloads. The EFK Stack (Elasticsearch, Fluentd, and Kibana) is a common logging solution in Kubernetes. **Fluentd** collects and forwards logs. **Elasticsearch** stores logs for querying and analysis [202]. **Kibana** visualizes logs with dashboards and search capabilities.

- Deploy Fluentd for Log Collection: Fluentd runs as a DaemonSet, ensuring logs are collected from every node in the cluster.
  - Define Fluentd DaemonSet (fluentd.yaml)

```
apiVersion: apps/v1
kind: DaemonSet
metadata:
  name: fluentd
spec:
  template:
    spec:
```

```
containers:
  - name: fluentd
    image: fluent/fluentd
```

```
kubectl apply -f fluentd.yaml
```

– Apply Fluentd Configuration: Deploy Fluentd to the cluster:

```
kubectl apply -f fluentd.yaml
```

This will run Fluentd on all nodes to aggregate logs.

- Deploy Kibana for Log Analysis: Kibana provides a web interface to analyze logs stored in Elasticsearch.

```
kubectl apply -f kibana.yaml
kubectl port-forward svc/kibana 5601:5601
```

Access Kibana at: http://localhost:5601

Kubernetes logging with the EFK Stack enhances observability by efficiently collecting, storing, and analyzing logs. Fluentd gathers logs, Elasticsearch indexes them, and Kibana provides a user-friendly interface for visualization and troubleshooting.

14.8 Best Practices

Following best practices in Kubernetes enhances security, reliability, and performance. Below are key recommendations to maintain a secure and efficient cluster.

- Run Containers as Non-Root Users: By default, containers run as the root user, which poses a security risk. Setting a security context ensures containers run with a non-root user. Example: Define a security context in a Pod spec

```
securityContext:
  runAsUser: 1000
  runAsGroup: 3000
  fsGroup: 2000
```

This configuration enforces running containers with a non-root user (1000), reducing the risk of privilege escalation.

- Apply Network Policies: Network policies restrict traffic flow between pods, enhancing security by preventing unauthorized access [188].

```yaml
apiVersion: networking.k8s.io/v1
kind: NetworkPolicy
metadata:
  name: deny-all
spec:
  podSelector: {}
  policyTypes:
  - Ingress
```

This policy denies all incoming traffic to pods. After apply the Network Policy

```
kubectl apply -f network-policy.yaml
```

This ensures that only explicitly allowed traffic can reach Kubernetes workloads.

Implementing best practices like running containers as non-root users and enforcing network policies significantly enhances security in Kubernetes. These measures prevent privilege escalation, unauthorized access, and mitigate potential threats to the cluster.

14.9 Case Study: Tesla Attack

In 2018, Tesla's Kubernetes cluster was compromised due to misconfigurations, leading to unauthorized use of cloud resources for cryptojacking [203]. The attackers exploited security vulnerabilities in Tesla's cloud infrastructure, specifically targeting an exposed Kubernetes Dashboard that lacked authentication and access controls. This allowed them to deploy crypto-mining containers, consuming Tesla's computing resources for illicit cryptocurrency mining. The root cause of the breach was the exposure of sensitive credentials and the absence of Role-Based Access Control (RBAC) restrictions, which left the cluster vulnerable to unauthorized access. To prevent similar attacks, organizations should implement several critical security measures. First, enabling authentication for the Kubernetes Dashboard is essential to prevent unauthorized access and ensure only authorized users can manage cluster resources. Second, implementing RBAC and network policies helps restrict access to critical resources and enforce the principle of least privilege, minimizing the risk of privilege escalation and unauthorized actions. Finally, continuous monitoring of cluster activity using tools like Prometheus and Grafana allows organizations to detect anomalies in real time and respond swiftly to potential threats. By adopting these security best practices, organizations can strengthen their Kubernetes environments and protect their cloud infrastructure from similar exploits. In conclusion, the Tesla Kubernetes attack highlights the risks of misconfigured cloud environments. Proper security controls, such as RBAC, authentication, and monitoring, are essential to prevent unauthorized access and protect cloud resources.

14.10 Security Tools

Security tools play a vital role in ensuring robust authentication and access control within Kubernetes environments. One such tool is Keycloak, an identity and access management solution that integrates seamlessly with Kubernetes to provide authentication services for users and applications [204].

- Keycloak: Keycloak allows organizations to manage user identities, enforce authentication policies, and integrate with various identity providers. To deploy Keycloak in Kubernetes, follow these steps: **Install Keycloak:**

```
kubectl create namespace keycloak
kubectl apply -f https://raw.githubusercontent.com/keycloak/
    keycloakoperator/
main/deploy/operator.yaml
```

  This installs the Keycloak Operator, which helps manage Keycloak instances within the Kubernetes cluster. **Access Keycloak Dashboard:**

```
kubectl port-forward svc/keycloak
    8080:8080
```

  Once the port-forwarding is enabled, users can access the Keycloak authentication dashboard by navigating to: http://localhost:8080/auth

Keycloak enhances Kubernetes security by providing centralized authentication and user management. Integrating Keycloak with Kubernetes helps enforce strong access controls, ensuring only authorized users can interact with cluster resources.

14.11 CI/CD Security

Continuous Integration and Continuous Deployment (CI/CD) pipelines streamline software delivery but also introduce security risks if not properly configured [205]. To ensure secure deployments in Kubernetes, organizations must incorporate security best practices such as image scanning, admission controls, and secret management [206].

- Scan Images in CI/CD Pipelines: Before deploying container images, it is essential to scan them for vulnerabilities using tools like Trivy, Clair, or Snyk. The following example demonstrates how to integrate Trivy into GitHub Actions for automated image scanning: **Example: Scan in GitHub Actions**

```
jobs:
  security-scan:
    runs-on: ubuntu-latest
    steps:
```

```
    - name: Checkout Code
      uses: actions/checkout@v2
    - name: Scan Image with Trivy
      run: trivy image myapp:latest
```

- Enable Admission Controllers: It help enforce security policies before workloads are scheduled in Kubernetes. To apply a Pod Security Policy [207], use the following command:

```
kubectl apply -f pod-security-policy.yaml
```

  This restricts insecure configurations and ensures compliance with security standards.
- Secure Kubernetes Secrets: Sensitive data such as database credentials should not be hardcoded in configuration files. Kubernetes Secrets provide a secure way to store and manage credentials.

```
kubectl create secret generic db-creds --from-
    literal=username=admin --from-literal=
    password=securepass
```

  This ensures that credentials are securely stored and only accessible to authorized applications.

Implementing security in CI/CD pipelines helps mitigate vulnerabilities before deployment. By scanning container images, enforcing admission controls, and securely managing secrets, organizations can significantly enhance the security of their Kubernetes-based CI/CD workflows.

14.12 Key Takeaways

- Role-Based Access Control (RBAC), authentication, and authorization ensure that only authorized users and services interact with the cluster, minimizing security risks.
- Monitoring tools like Prometheus and Grafana provide real-time observability, helping administrators detect performance issues and security threats.
- Logging solutions such as Fluentd and Elasticsearch enhance traceability by capturing critical events, aiding in debugging and compliance audits.
- Best security practices, including enforcing network policies, securing container images, implementing the principle of least privilege, and regularly updating security configurations, strengthen Kubernetes security.
- CI/CD security measures prevent vulnerabilities from being introduced during automated software deployment, ensuring a secure development pipeline.

- The Tesla Attack case study underscores the importance of proactive security measures, proper access controls, and continuous monitoring to prevent unauthorized access and data breaches.
- A comprehensive security and monitoring strategy ensures a resilient Kubernetes environment, protecting workloads from evolving threats and misconfigurations.

14.13 Summary

This chapter provided an in-depth exploration of Kubernetes security and monitoring, covering fundamental concepts, best practices, and real-world challenges. It began by explaining critical security components such as authentication, authorization, and Role-Based Access Control (RBAC), which collectively ensure controlled and secure access to Kubernetes clusters. The chapter then introduced monitoring solutions like Prometheus, Grafana, and Kubernetes Dashboard, highlighting their role in maintaining cluster health and providing real-time insights into workloads. Logging and observability were also discussed, with a focus on tools such as Fluentd and Elasticsearch, which facilitate centralized log management and security auditing. Best practices for Kubernetes security were outlined, including implementing network policies to restrict unauthorized communication, securing container images against vulnerabilities, enforcing the least privilege principle to minimize exposure, and regularly patching system components. CI/CD security was also emphasized, as insecure deployment pipelines can introduce risks that may compromise container workloads and sensitive data. A critical case study on the Tesla Attack illustrated the impact of misconfigurations and poor security practices, reinforcing the need for proactive security monitoring and robust access controls. The chapter concluded by stressing the importance of a multi-layered security approach, integrating monitoring and security tools, following industry best practices, and staying vigilant against evolving threats to ensure a secure and resilient Kubernetes ecosystem.

14.14 Review Questions

- What are the core components of Kubernetes security architecture?
- How does Role-Based Access Control (RBAC) enhance Kubernetes security?
- What are the key differences between authentication and authorization in Kubernetes?
- What security risks does the Kubernetes Dashboard present, and how can they be mitigated?
- How can you implement network policies to improve cluster security?
- What lessons can be learned from the Tesla Kubernetes attack case study?

- Which open-source security tools are most effective for Kubernetes environments?
- How does Pod Security Admission enhance workload protection?
- What CI/CD security measures should be implemented for Kubernetes deployments?
- How can you monitor etcd for security and performance issues?
- Which built-in monitoring tools come with Kubernetes, and what do they track?
- How does Prometheus integrate with Kubernetes for monitoring?
- What are the best practices for logging in Kubernetes clusters?
- What are the three pillars of Kubernetes observability (metrics, logs, traces)?
- How do service meshes like Istio contribute to Kubernetes security?

Chapter 15
Advanced Topics, Kubernetes Certification and Career

Abstract This chapter explores advanced Kubernetes topics, practical deployment strategies, and career pathways. It begins with CI/CD integration and the use of operators to automate complex operational tasks. Real-world case studies, including the migrations of Pokémon Go and Booking.com, illustrate large-scale Kubernetes adoption and best practices. High availability through multi-master setups and package management with Helm charts are detailed, alongside troubleshooting strategies to resolve common cluster issues. Future trends in Kubernetes development and ecosystem evolution are highlighted to inform forward-looking strategies. The chapter also addresses career opportunities, offering guidance on certification programs, training resources, and common interview questions across beginner, intermediate, and advanced levels. Emphasis is placed on mastering Kubernetes through hands-on practice and community engagement, providing readers with the skills and knowledge required to excel in professional roles within cloud-native environments.

Keywords CI/CD integration · Operators · Multi-master Setup · Helm charts · Troubleshooting · Certification

This chapter covers advanced Kubernetes topics, including CI/CD integration, operators, high availability with multi-master setups, and Helm charts for package management. It also explores real-world case studies, such as Pokémon Go and Booking.com migrations, to highlight large-scale Kubernetes deployments. Additionally, troubleshooting techniques and future trends in Kubernetes are discussed. The second part of the chapter focuses on Kubernetes certification pathways, training resources, and career development. It provides a structured list of common interview questions, ranging from beginner to advanced levels, and emphasizes hands-on practice and community engagement as key strategies for mastering Kubernetes. Kubernetes Federation enables the management of multiple clusters as a unified system, allowing administrators to deploy applications and enforce policies across multiple Kubernetes clusters from a single control plane [208]. This approach provides several key benefits: **Centralized Deployment and Policy**

B. Kumar et al., *Modern Kubernetes: From Core Concepts to Intelligent Autoscaling for Cloud Applications*, Studies in Autonomic, Data-driven and Industrial Computing, https://doi.org/10.1007/978-3-032-12972-7_15

Management: It ensures consistency across clusters by applying policies and configurations uniformly [209]. **Enhanced Disaster Recovery:** Supports multi-cluster failover, ensuring applications remain available even if a cluster experiences issues. **Improved Availability and Load Balancing:** It distributes workloads dynamically across clusters, optimizing resource usage and performance.

Federation Setup (KubeFed v2)

- Install KubeFed: This involves setting up the federation namespace and applying the latest KubeFed manifest [210].

```
kubectl create ns kube-federation-system
kubectl apply -f https://github.com/kubernetes-sigs/kubefed/
    releases/latest/download/kubefed.yaml
```

- Join a Cluster to Federation: The kubefedctl join command registers a cluster to the federation.

```
kubefedctl join cluster1 --host-cluster-context=
    main-context
```

- Deploy Workloads Across Clusters: Federated Deployments ensure applications are replicated across clusters for high availability. The given command is a Kubernetes Federation resource definition for deploying an application across multiple clusters using Kubefed [211], the Kubernetes Federation tool. Let's break down each component of the YAML configuration:

```
apiVersion: types.kubefed.io/v1beta1
kind: FederatedDeployment
metadata:
  name: nginx
spec:
  template:
    spec:
      replicas: 3
```

The apiVersion: types.kubefed.io/v1beta1 specifies the API version of Kubernetes Federation, commonly known as Kubefed. The v1beta1 version indicates that this feature is still in the beta stage, meaning it is functional but may undergo future improvements or changes. The kind: FederatedDeployment defines the type of Kubernetes object being created. A FederatedDeployment ensures that the deployment configuration is consistently applied across multiple federated clusters, allowing centralized management and coordination of workloads. This helps maintain uniformity and availability in distributed environments. The metadata section provides essential information about the deployment. The name: nginx assigns the name nginx to the federated deployment, making it easily identifiable. This name helps Kubernetes and administrators manage and track the deployment across different clusters. Kubernetes Federation offers several key use cases that enhance the management and deployment of workloads across multiple clusters.

Multi-region deployments allow organizations operating in different geographical locations to ensure low-latency access by deploying applications closer to users, improving performance and responsiveness. High availability and failover capabilities enable automatic shifting of critical applications to healthy clusters in case of failures, minimizing downtime and ensuring continuous service. Compliance and governance are strengthened through centralized policy enforcement, ensuring regulatory compliance across all clusters. Additionally, cross-cloud deployments benefit enterprises using hybrid or multi-cloud strategies, allowing seamless workload management across different cloud providers, optimizing resource utilization, and reducing vendor lock-in. By streamlining multi-cluster management, Kubernetes Federation enables centralized control, enhances resilience through failover mechanisms, and optimizes workload distribution. It serves as a powerful solution for enterprises seeking scalability, high availability, and operational efficiency in distributed Kubernetes environments.

15.1 CI/CD Integration

Continuous Integration and Continuous Deployment (CI/CD) pipelines are crucial for automating application deployment in Kubernetes environments. By integrating CI/CD workflows, developers can streamline application updates, reduce manual intervention, and ensure rapid, reliable releases. Kubernetes provides various tools and strategies for implementing CI/CD, allowing automated deployment, testing, and monitoring of applications across clusters.

One commonly used method for CI/CD integration in Kubernetes is through GitHub Actions, which enables seamless automation of deployment workflows. The following YAML configuration illustrates a simple GitHub Actions CI/CD pipeline for Kubernetes deployment:

```
jobs:
  deploy:
    runs-on: ubuntu-latest
    steps:
    - name: Checkout Code
      uses: actions/checkout@v2
    - name: Set up Kubectl
      uses: azure/setup-kubectl@v1
    - name: Deploy to Kubernetes
      run: kubectl apply -f deployment.yaml
```

The job definition (deploy) in the CI/CD pipeline is responsible for deploying an application to a Kubernetes cluster. This job is executed in an Ubuntu-based environment (runs-on: ubuntu-latest), ensuring compatibility with most Kubernetes deployment tools. It provides a reliable and standardized environment for running deployment tasks efficiently. The checkout code step (actions/checkout@v2) retrieves the source code from the repository, ensuring that all necessary deployment files are available in the pipeline. This step is essential as it allows the pipeline to access and deploy the latest version of the application, keeping deployments up to date. Next, the setup for kubectl (azure/setup-kubectl@v1) installs and configures kubectl, the Kubernetes command-line tool. This tool is crucial for interacting with the cluster, enabling the pipeline to apply configurations, manage resources, and monitor the deployment process. Without this step, the pipeline would be unable to communicate with the Kubernetes cluster. The final step, deploying to Kubernetes (kubectl apply -f deployment.yaml), executes the deployment process by applying the deployment.yaml file. This file contains the configuration details necessary to create or update Kubernetes resources, such as pods, services, and other workload components. By running this command, the pipeline ensures that the application is deployed or updated based on the latest configuration settings. While GitHub Actions is a widely used CI/CD solution, several alternative tools provide advanced Kubernetes deployment capabilities. ArgoCD is a GitOps-based continuous delivery tool that maintains Kubernetes resources in a desired state by syncing with a Git repository [212]. Similarly, FluxCD is another GitOps tool that automates deployments and updates based on repository changes [213]. For organizations looking for a Kubernetes-native CI/CD framework, Tekton offers a flexible and scalable solution for automating pipeline execution. These tools enhance Kubernetes deployment processes, providing greater flexibility, automation, and operational efficiency.

In conclusion, CI/CD integration in Kubernetes automates the deployment and management of applications, ensuring efficient and error-free releases. The provided GitHub Actions example demonstrates a simple yet effective deployment pipeline using kubectl. Alternative tools like ArgoCD, FluxCD, and Tekton provide additional flexibility, especially for organizations adopting GitOps methodologies. By leveraging Kubernetes CI/CD pipelines, teams can improve deployment speed, reduce downtime, and maintain consistent application updates across distributed environments.

15.2 Operators

Operators extend Kubernetes functionality by automating application-specific tasks, simplifying the management of complex workloads [214]. They encapsulate human operational knowledge into software, enabling Kubernetes to manage applications efficiently. Operators are particularly useful for automating lifecycle management tasks such as provisioning, scaling, and upgrades for stateful applications. Example: Creating a Custom Operator: To create a Kubernetes Operator, the Operator SDK provides a framework that streamlines the development process. The following steps demonstrate how to set up a custom operator:

- Install Operator SDK: The Operator SDK must be installed before creating an operator. The following commands download and install the Operator SDK on a Linux system:

```
curl -sLO
 https://github.com/operator-framework/operator-sdk/releases/
     latest/
download/operator-sdk_linux_amd64
chmod +x operator-sdk
mv operator-sdk /usr/local/bin/
```

These commands fetch the latest release, provide executable permissions, and move the binary to a system-wide location for easier access.

- Generate Operator Scaffolding: After installing the SDK, the next step is to initialize the operator project structure using the following command:

```
            operator-sdk init --domain=mydomain.com --repo=
                github.com/example/my-operator
```

This command sets up the necessary scaffolding for an operator, including configuration files, a controller, and CRD (Custom Resource Definition) templates. By using Kubernetes Operators, developers can extend the cluster's capabilities beyond built-in controllers, enabling custom automation and self-healing mechanisms for complex applications.

15.3 Case Study: Pokémon Go and Booking.com Migration

- Pokémon Go: Pokémon Go experienced an unprecedented surge in traffic, scaling from zero to over 50 million users within days [215]. The rapid growth overwhelmed game servers, causing performance issues and service disruptions. To address these challenges, the development team leveraged Kubernetes on Google Cloud Platform (GCP). By implementing autoscaling, multi-cluster load balancing, and global traffic distribution through Google Cloud Load Balancer, Pokémon Go ensured a seamless gaming experience despite the massive traffic influx. Kubernetes played a critical role in dynamically allocating resources, improving system resilience, and maintaining high availability.
  - Challenges and Solutions: Pokémon Go faced a significant challenge when it rapidly scaled from zero to over 50 million users within days, causing massive traffic surges that overloaded game servers. To address this, the developers adopted Kubernetes on Google Cloud Platform (GCP), leveraging its autoscaling capabilities, load balancing across multiple clusters, and global traffic distribution via Google Cloud Load Balancer. This approach ensured seamless scalability and efficient resource management, enabling the game to handle its unprecedented user growth.

- Booking.com Migration: Booking.com faced the challenge of migrating from OpenShift to a self-managed Kubernetes environment while handling millions of travel search queries [216]. The transition required a scalable and flexible infrastructure that could support its complex workloads. To achieve this, the company built a custom Kubernetes platform, incorporating Helm charts and Operators for automation. Additionally, they implemented CI/CD pipelines to accelerate deployments and streamline application updates. This migration enhanced operational efficiency, improved resource utilization, and provided greater control over the infrastructure, enabling Booking.com to meet growing business demands effectively.
  - Challenges and Solutions: Migrating from OpenShift to a self-managed Kubernetes environment posed significant challenges, particularly in handling large-scale travel search queries efficiently. To overcome these issues, the team built a custom Kubernetes platform, leveraging Helm and Operators for automation and implementing a CI/CD pipeline to enable fast and reliable deployments. This approach streamlined operations, improved scalability, and enhanced deployment efficiency.

Both case studies highlight Kubernetes' ability to support large-scale applications by offering scalability, automation, and resilience in highly demanding environments.

15.4 Multi-master Setup (High Availability)

A multi-master Kubernetes setup is essential for ensuring high availability, fault tolerance, and scalability in production environments [217]. By deploying multiple master nodes, organizations can prevent a single point of failure, ensure zero-downtime upgrades, and efficiently distribute API requests across the cluster.

- Why Multi-Master: A multi-master setup is essential for preventing a single point of failure, ensuring that if one master node fails, others can continue managing the cluster. It also enables zero-downtime upgrades by allowing rolling updates without disrupting cluster availability. Additionally, multi-master architecture efficiently distributes API requests through load balancing, enhancing responsiveness and overall system performance.
- Setup Multi-Master Kubernetes Cluster
  - Deploy an Etcd Cluster for High Availability: The etcd cluster serves as the distributed key-value store for Kubernetes, ensuring consistency and reliability across master nodes. Initialize the cluster with:

```
kubeadm init --control-plane-endpoint
 "load-balancer-ip:6443"
```

- Join Additional Masters Nodes: Additional master nodes are joined to the cluster using the following command:
  * kubeadm join load-balancer-ip:6443 –token xyz –discovery-token-ca-cert-hash sha256:abc

A multi-master Kubernetes setup enhances system resilience, allowing uninterrupted cluster operation even when individual master nodes fail. This architecture is crucial for mission-critical applications requiring continuous availability and reliability.

15.5 Helm Charts (Package Management)

Helm is a package manager for Kubernetes that simplifies application deployment, version management, and configuration [218]. It enables developers to define, install, and upgrade applications seamlessly using Helm charts, which bundle Kubernetes resources into reusable templates.

- Install Helm: To install Helm, use the following command:

```
curl https://raw.githubusercontent.com/helm/helm/main/scripts
    /get-helm-3 | bash
```

  This script downloads and installs Helm version 3, which eliminates the need for a server-side component (Tiller) and improves security and usability.
- Deploy an App with Helm: Helm allows quick application deployment using predefined charts. For example, to deploy an NGINX application:

```
            helm repo add stable https://charts.helm.sh/
                stable
            helm install myapp stable/nginx
```

  The first command adds the stable Helm chart repository. The second command deploys an NGINX application using the stable/nginx chart and assigns it the name myapp.

Helm streamlines application management by enabling simple rollbacks, version control, and automated configuration, making it an essential tool for Kubernetes administrators and developers.

15.6 Troubleshooting Kubernetes

Troubleshooting Kubernetes [170] involves diagnosing and resolving common issues that affect pod scheduling, application performance, networking, and DNS resolution. Below are key challenges and solutions:

- Common Issues and Solutions
  - Pod Stuck in Pending: A pod may remain in a Pending state due to insufficient resources, scheduling constraints, or missing dependencies. Use the following command to check the pod's status and scheduling events:

    ```
    kubectl describe pod mypod
    ```

    This command provides details on node availability, resource requests, and scheduling failures, helping to identify the root cause.
- Check node availability
  - CrashLoopBackOff Error: A pod in a CrashLoopBackOff state continuously restarts due to errors in the application or missing dependencies. To check logs for failure reasons, run:

    ```
    kubectl logs mypod
    ```

    This helps diagnose application crashes, misconfigurations, or missing environment variables.
- Check application logs for failures
  - Network Connectivity Issues: If a service is not reachable, use the following command to verify its status and associated pods:

    ```
    kubectl get svc -o wide
    ```

    This command helps inspect service endpoints, IP assignments, and connectivity issues between pods.
- Verify service and pod network configuration
  - DNS Resolution Failures: If a pod cannot resolve domain names, it may indicate a CoreDNS issue. Test DNS resolution by executing:

    ```
    kubectl exec -it mypod -- nslookup google.com
    ```

If DNS failures persist, restarting CoreDNS may resolve the issue:

```
kubectl delete pod -n kube-system -l k8s-app=
    kube-dns
```

This ensures that CoreDNS restarts with fresh configurations.

- Restart CoreDNS if needed

By following these troubleshooting steps, Kubernetes administrators can quickly diagnose and resolve common cluster issues, ensuring high availability and smooth application operations.

15.7 Future Trends

The future of Kubernetes is evolving with innovations in serverless computing and AI-driven workloads, enabling greater efficiency, automation, and scalability. Below are two key trends shaping Kubernetes advancements:

- Serverless Kubernetes (Knative): Knative is a serverless framework designed to run on Kubernetes, allowing applications to scale dynamically based on demand [208]. It simplifies event-driven execution, automatic scaling (including scaling to zero), and reduces infrastructure costs. To deploy a Knative service, use the following YAML configuration:

```
apiVersion: serving.knative.dev/v1
kind: Service
metadata:
  name: knative-app
spec:
  template:
    spec:
      containers:
      - image: gcr.io/knative-samples/helloworld-go
```

  Apply the service using:

```
kubectl apply -f knative-service.yaml
```

  Key Benefits: **Automatic scaling**, including the ability to scale to zero, dynamically adjusts resources based on workload demand, minimizing resource waste. **Event-driven** execution enables real-time responsiveness, ensuring efficient workload management by scaling down unused resources to zero, ultimately optimizing cloud expenses. Additionally, this approach leads to **lower infrastructure costs** by reducing unnecessary resource consumption.
- Kubernetes AI Workloads: Kubernetes is increasingly being utilized for machine learning (ML) and artificial intelligence (AI) workloads, providing robust

orchestration, model training, and inference capabilities [219]. **Kubeflow**, a Kubernetes-native ML framework, enables efficient orchestration of ML pipelines, distributed training, and AI model deployment [220]. Additionally, **KServe** offers a scalable solution for serving AI/ML models in production, supporting multiple frameworks such as TensorFlow, PyTorch, and ONNX, ensuring seamless integration and scalability for AI applications.

These advancements position Kubernetes as a powerful AI and serverless computing platform, making it a key player in the future of cloud-native applications.

15.8 Kubernetes Certification and Career

Kubernetes has become the cornerstone of modern cloud-native applications, driving demand for skilled professionals who can manage and optimize containerized environments [221]. The Certified Kubernetes Administrator (CKA) certification serves as a benchmark for validating expertise in Kubernetes cluster management, security, networking, and troubleshooting [222]. This chapter explores the CKA certification in detail, including preparation strategies, exam structure, prerequisites, career benefits, training resources, interview questions, and practical exercises [15]. Additionally, it provides guidance on leveraging community networks for career advancement [223].

- The CKA Certification Overview: The Certified Kubernetes Administrator (CKA) certification is an industry-recognized credential that demonstrates proficiency in managing Kubernetes clusters. It is designed for DevOps engineers, Site Reliability Engineers (SREs), and cloud professionals who want to establish themselves as Kubernetes experts. The exam assesses the candidate's ability to set up, administer, and troubleshoot Kubernetes clusters effectively. This certification covers crucial areas such as cluster installation and configuration, networking, security, workload and storage management, and overall system troubleshooting. Unlike multiple-choice exams, CKA is a hands-on, scenario-based test where candidates perform real-time tasks on a Kubernetes cluster, ensuring practical expertise in dealing with real-world operational challenges.
- Preparing for the CKA Exam: To successfully clear the CKA exam, candidates must focus on hands-on practice rather than just theoretical knowledge. Setting up a Minikube environment or working with cloud-based Kubernetes playgrounds such as KodeKloud Labs, Play with Kubernetes (by Docker), or Kubernetes the Hard Way (by Kelsey Hightower) provides valuable experience. These environments help candidates practice essential commands such as deploying applications, configuring networking, managing storage, and troubleshooting cluster issues. Mastery of kubectl commands, YAML configurations, and Kubernetes architecture is crucial for effectively handling exam scenarios.

- Exam Structure and Scoring Criteria: The CKA exam is a proctored, browser-based test consisting of 17–20 practical tasks to be completed within two hours. Candidates are evaluated on their ability to perform administrative tasks on a live Kubernetes cluster. The test follows the latest Kubernetes v1.26+ version, ensuring relevance to current industry standards. The exam covers several core domains, with different weightages assigned to each:
  - Cluster Architecture, Installation and Configuration (25%)—Includes setting up clusters, configuring control planes, and managing API access.
  - Workloads and Scheduling (15%)—Focuses on deploying and scaling workloads using Deployments, DaemonSets, and Jobs.
  - Services and Networking (20%)—Involves configuring network policies, Services (ClusterIP, NodePort, LoadBalancer), and Ingress controllers.
  - Storage (10%)—Covers PersistentVolumes, PersistentVolumeClaims, and StorageClasses.
  - Troubleshooting (30%)—Evaluates the ability to diagnose and resolve issues such as node failures, network disruptions, and CrashLoopBackOff errors.

  The passing score for the CKA exam is 66%, and the performance is assessed based on the accuracy and efficiency of task execution.

15.8.1 Certification and Training Resources

Gaining expertise in Kubernetes requires a combination of certification and hands-on training. The right resources can help candidates develop a strong foundation in Kubernetes concepts, prepare for certification exams, and enhance their practical skills for real-world deployments

- Prerequisites for the CKA Certification [15]: While there are no formal prerequisites for taking the CKA exam, candidates should have a strong foundation in Linux and containerization concepts [224]. Familiarity with fundamental Linux commands such as ls, cd, vim, curl, chmod, and systemctl is essential. Understanding YAML configurations is equally important, as Kubernetes resource definitions rely heavily on YAML syntax. Networking knowledge, particularly Pod-to-Pod communication, Service types (ClusterIP, NodePort, LoadBalancer), and DNS resolution, is also a key requirement.
- Career Benefits of CKA Certification: Earning the CKA certification significantly enhances career prospects in the cloud and DevOps space. CKA-certified professionals are in high demand worldwide, with salaries ranging from 120K to 170K per year, depending on experience and geographical location. The certification is backed by the Cloud Native Computing Foundation (CNCF), ensuring credibility and recognition across industries. Common job roles for CKA-certified professionals include Kubernetes Administrator, Site Reliability Engineer (SRE), and DevOps Engineer. Many organizations specifically list CKA certification as a

preferred qualification for hiring Kubernetes professionals, making it a valuable asset for career growth.

- Recommended Training Resources: Several training resources are available to help candidates prepare for the CKA exam. The official Kubernetes documentation (kubernetes.io) remains the best free reference for in-depth knowledge. Hands-on courses such as KodeKloud CKA Course, Udemy's CKA Exam Prep, and Edureka's Kubernetes Training offer structured learning paths with practical labs. Books such as "Kubernetes Up and Running" by Brendan Burns and "The Kubernetes Book" by Nigel Poulton provide valuable insights into Kubernetes fundamentals and advanced concepts [13]. These resources, combined with regular hands-on practice, ensure comprehensive preparation for the exam.

15.9 Common Interview Questions for Kubernetes Roles

Preparing for Kubernetes-related job interviews is essential for securing roles in DevOps, cloud engineering, and site reliability [225]. Employers assess both foundational and advanced Kubernetes knowledge to gauge a candidate's expertise in managing containerized workloads. Apart from obtaining certifications, candidates should also prepare for **Kubernetes-related job interviews**, as employers assess both fundamental and advanced knowledge through various questions. Below is a categorized list of common interview topics:

15.9.1 Beginner-Level Questions

These questions focus on Kubernetes basics, core components, and fundamental operations:

- *What is Kubernetes, and why is it used?*
- *What is a Pod in Kubernetes?*
- *How does a Service expose applications?*
- *What is a ConfigMap and how is it used?*
- *What is the role of a kubelet in a Kubernetes cluster?*
- *How does a ReplicaSet ensure high availability?*
- *What is the difference between a Deployment and a StatefulSet?*

15.9.2 Intermediate-Level Questions

These questions cover networking, scaling, security, and storage in Kubernetes:

- *How does an Ingress controller manage traffic in a Kubernetes cluster?*
- *What are the different ways to scale a Kubernetes deployment?*
- *What are PersistentVolumes (PV) and PersistentVolumeClaims (PVC)?*
- *How does Kubernetes handle stateful applications with StatefulSets?*
- *What are the differences between a DaemonSet and a Deployment?*
- *How does Kubernetes implement Role-Based Access Control (RBAC)?*

15.9.3 Advanced-Level Questions

These questions assess deep technical expertise in troubleshooting, optimization, and Kubernetes internals:

- *How would you troubleshoot a Pod stuck in a CrashLoopBackOff state?*
- *What are the advantages and limitations of Horizontal Pod Autoscaler (HPA) and Vertical Pod Autoscaler (VPA)?*
- *How does Kubernetes ensure high availability (HA) and fault tolerance?*
- *How does Kubernetes handle resource requests, limits, and scheduling?*
- *What strategies can be used for rolling updates and rollbacks in Kubernetes?*
- *What are custom controllers and operators, and how do they extend Kubernetes functionality?*

Being well-prepared for these questions ensures confidence in **technical interviews** and enhances job prospects in Kubernetes-related roles.

15.10 Mastering Kubernetes Through Practice and Community

Mastering Kubernetes requires a combination of hands-on experience, structured learning, and active engagement with the community. Practical exercises such as deploying an Nginx Pod using YAML, configuring ConfigMaps and Secrets, scaling deployments with Horizontal and Vertical Pod Autoscalers, and implementing StatefulSets for stateful applications help build expertise in managing workloads. Advanced tasks like setting up a multi-node Kubernetes cluster, managing network policies, and securing containerized applications with Role-Based Access Control (RBAC) and Pod Security Policies deepen practical knowledge. Beyond individual practice, learning from the broader Kubernetes community accelerates growth. Platforms like CNCF Slack Channels, GitHub, and Reddit's /r/kubernetes serve as hubs for troubleshooting, best practices, and collaboration. Attending industry events like KubeCon provides opportunities to engage with experts, discover emerging trends, and gain insights into production-grade Kubernetes deployments. Contributing to open-source projects on GitHub not only strengthens technical skills but also enhances professional credibility and career prospects. By actively participating in discussions, sharing solutions, and working on community-driven projects, candidates gain real-world experience and establish themselves as skilled Kubernetes practitioners.

1. Hands-on Practice is Essential for Mastering Kubernetes
   - Deploy an Nginx pod using YAML:

```
apiVersion: v1
kind: Pod
metadata:
  name: nginx-pod
spec:
  containers:
    - name: nginx
      image: nginx:latest
```

   - Apply the configuration with:

```
kubectl apply -f nginx-pod.yaml
```

2. Using ConfigMaps and Secrets for Configuration Management
   - Create a ConfigMap:

```
kubectl create configmap app-config --from-
    literal=ENV=production
```

   - Create a Secret:

```
kubectl create secret generic db-secret
    --from-literal=password=
    mysecurepassword
```

3. Scaling Deployments Dynamically
   - Scale an Application using the Horizontal Pod Autoscaler (HPA):

```
kubectl autoscale deployment my-app --cpu-
    percent=50 --min=1 --max=10
```

4. Managing Stateful applications with StatefulSets
   - Deploy a StatefulSet:

```
apiVersion: apps/v1
kind: StatefulSet
metadata:
  name: web
spec:
  serviceName: "web"
  replicas: 3
  selector:
    matchLabels:
      app: web
  template:
    metadata:
      labels:
        app: web
    spec:
      containers:
      - name: web
        image: nginx
```

5. Advanced Kubernetes Networking and Security
   - Apply a network policy to restrict pod communication:

```
apiVersion: networking.k8s.io/v1
kind: NetworkPolicy
metadata:
  name: deny-all
spec:
  podSelector: {}
  policyTypes:
  - Ingress
```

   - Apply Role-Based Access Control (RBAC):

```
apiVersion: rbac.authorization.k8s.io/v1
kind: Role
metadata:
  namespace: default
  name: pod-reader
rules:
- apiGroups: [""]
  resources: ["pods"]
  verbs: ["get", "watch", "list"]
```

6. Leveraging Community Resources for Learning and Troubleshooting
   - Join CNCF Slack (https://slack.cncf.io/)
   - Follow Kubernetes GitHub discussions (https://github.com/kubernetes/kubernetes/discussions)
   - Participate in Kubernetes-related subreddits (https://www.reddit.com/r/kubernetes/)
   - Attend KubeCon (https://events.linuxfoundation.org/kubecon-cloudnativecon/)
7. Contributing to Open-Source Projects on GitHub
   - Fork a Kubernetes-related repository and make contributions:

```
git clone https://github.com/kubernetes/kubernetes.git
cd kubernetes
git checkout -b feature-branch
git push origin feature-branch
```

By combining hands-on exercises, security best practices, scaling techniques, and active community involvement, professionals can build deep expertise and advance their careers in Kubernetes.

15.11 Key Takeaways

- CI/CD integration enhances automation and security by ensuring efficient and secure application deployments, minimizing risks from insecure pipelines.
- Kubernetes Operators streamline complex application management by automating deployment, scaling, and maintenance tasks, making it easier to manage stateful applications.
- Real-world case studies, such as Pokémon Go and Booking.com migration, demonstrate how Kubernetes enables scalability, reliability, and efficient resource management in large-scale applications.
- A multi-master setup ensures high availability by distributing the control plane across multiple nodes, improving resilience and fault tolerance in production environments.
- Helm charts simplify Kubernetes package management, allowing developers to deploy and manage applications consistently across different environments.
- Effective troubleshooting strategies help resolve common Kubernetes issues, such as networking failures, misconfigurations, and resource constraints, ensuring system stability.
- Future trends in Kubernetes include AI-driven automation, serverless computing, and enhanced security mechanisms, shaping the evolution of cloud-native ecosystems.

- Kubernetes certification and training resources help professionals validate their skills, with industry-recognized certifications like CKA (Certified Kubernetes Administrator) and CKAD (Certified Kubernetes Application Developer).
- Common Kubernetes interview questions range from beginner to advanced levels, covering core concepts, practical troubleshooting, and best practices for managing Kubernetes environments.
- Mastering Kubernetes requires hands-on experience and active community participation, including open-source contributions, real-world projects, and continuous learning through industry events.

15.12 Summary

Kubernetes continues to evolve as a powerful and resilient platform for managing cloud-native applications, offering automation, scalability, and reliability. Key advancements include seamless CI/CD integration using tools like GitHub Actions, ArgoCD, and Tekton, enabling efficient and automated deployments. Operators enhance Kubernetes' functionality by automating complex application-specific tasks, reducing manual intervention, and improving deployment efficiency. Real-world case studies, such as Pokémon Go and Booking.com, demonstrate Kubernetes' ability to handle large-scale workloads through autoscaling, load balancing, and high availability mechanisms. Multi-master setups ensure fault tolerance and zero-downtime upgrades, while Helm charts simplify package management, streamlining application deployment across diverse environments. Troubleshooting capabilities help resolve common issues, such as pod failures, network connectivity problems, and DNS errors, ensuring system stability. Emerging trends, including serverless computing with Knative and AI workload management with Kubeflow and KServe, further expand Kubernetes' role in modern cloud computing. For professionals looking to advance their careers in cloud-native technologies, Kubernetes certifications provide a valuable credential. The Certified Kubernetes Administrator (CKA) certification validates expertise in cluster management, security, and troubleshooting, making candidates highly desirable in the job market. Those interested in Kubernetes application development can pursue the Certified Kubernetes Application Developer (CKAD) certification, which focuses on designing, building, and deploying containerized applications. Success in these certifications requires hands-on practice, proficiency in kubectl commands, and strong troubleshooting skills. Additionally, active participation in the Kubernetes community–through forums, open-source contributions, and networking events like KubeCon–can further enhance career opportunities. As Kubernetes adoption grows across industries, professionals who continuously develop their skills and stay updated with emerging innovations will be well-positioned for leadership roles in the ever-evolving cloud ecosystem.

15.13 Review Questions

- How does CI/CD integration improve Kubernetes deployment workflows?
- What are Kubernetes Operators and how do they extend cluster functionality?
- What key lessons emerged from Pokémon Go's Kubernetes implementation?
- How did Booking.com benefit from migrating to Kubernetes?
- What are the essential components of a high-availability, multi-master Kubernetes setup?
- How do Helm charts simplify application deployment in Kubernetes?
- What systematic approach should be used for troubleshooting complex Kubernetes issues?
- Which official Kubernetes certifications are most valuable for career advancement?
- What are the best training resources for preparing for CKAD/CKA exams?
- What beginner-level concepts are commonly tested in Kubernetes interviews?
- What intermediate-level scenarios should candidates prepare for Kubernetes roles?
- How would you answer advanced questions about custom controllers or cluster networking?
- What hands-on practice methods most effectively build Kubernetes expertise?
- How can participating in the Kubernetes community accelerate professional growth?

References

1. Burns B, Beda J, Hightower K, et al (2016) Borg, omega, and Kubernetes. In: ACM queue, vol 14. 1. Google's paper on Kubernetes' origins and design principles. ACM, pp 70–93. https://queue.acm.org/detail.cfm?id=2898444
2. Cloud Native Computing Foundation (2021) CNCF Annual Survey 2021. Survey on Kubernetes adoption and cloud-native technologies. https://www.cncf.io/reports/cncf-annual-survey-2021/
3. Newman S (2020) Microservices with Kubernetes. In: IEEE software 37.3 . Best practices for microservices on Kubernetes, pp 7–11
4. Kubernetes Authors (2023) Kubernetes documentation. Official documentation for Kubernetes concepts and features. https://kubernetes.io/docs/home/
5. GitHub Engineering (2020) GitHub's journey to Kubernetes. Case study on GitHub's CI/CD pipeline with Kubernetes. https://github.blog/2020-09-25-githubs-journey-to-kubernetes/
6. Walmart Labs (2019) Walmart's cloud-native transformation. Tech. rep. Walmart's use of Kubernetes for Black Friday scalability. https://medium.com/walmartlabs/walmarts-cloud-native-journey-8f7c9b5c3e3a
7. Niantic Engineering (2016) Scaling Pokémon Go with Kubernetes. Case study on Kubernetes for real-time game scaling. https://nianticlabs.com/blog/kubernetes/
8. Kim G, Humble J, Debois P (2016) The DevOps handbook. Foundational principles of DevOps and CI/CD
9. Prometheus Community (2023) Observability in Kubernetes: logs, metrics, and traces. Tech. rep. Monitoring workflows with Prometheus and Grafana. https://prometheus.io/docs/introduction/overview/
10. Weaveworks (2021) GitOps: Kubernetes operations via Git. Introduction to GitOps methodology. https://www.weave.works/technologies/gitops/
11. Kubernetes Community (2023) kubectl Cheat Sheet. Official kubectl command reference. https://kubernetes.io/docs/reference/kubectl/cheatsheet/
12. Kubernetes Community (2023) Minikube documentation. Guide to local Kubernetes development with Minikube. https://minikube.sigs.k8s.io/docs/
13. Burns B, Beda J, Hightower K (2022) Kubernetes: up and running, 3rd edn. Comprehensive guide to Kubernetes concepts and practices. O'Reilly, Santa Rosa
14. Payscale (2023) DevOps and Kubernetes salary report. Salary trends for Kubernetes and DevOps roles. https://www.payscale.com/research/US/Job=DevOps_Engineer/Salary

B. Kumar et al., *Modern Kubernetes: From Core Concepts to Intelligent Autoscaling for Cloud Applications*, Studies in Autonomic, Data-driven and Industrial Computing, https://doi.org/10.1007/978-3-032-12972-7

15. Linux Foundation (2023) CKA certification study guide. Official guide for Certified Kubernetes Administrator (CKA) exam. https://training.linuxfoundation.org/certification/certified-kubernetes-administrator-cka/
16. Ibryam B, Huß R (2019) Kubernetes patterns: reusable elements for designing cloud-native applications. Design patterns for pods, deployments, and services in Kubernetes. O'Reilly, Santa Rosa
17. Red Hat OpenShift Team (2022) Kubernetes pod lifecycle and best practices. In: Detailed guide on Pod states and management. https://cloud.redhat.com/blog/kubernetes-pod-lifecycle
18. Doe J et al (2021) Sidecar and adapter patterns in Kubernetes pods. In: USENIX conference on operational Kubernetes. Design patterns for multi-container Pods. USENIX, pp 88–102
19. Kubernetes SIG-Architecture (2023) Kubernetes API conventions. Guidelines for Kubernetes YAML and API design. https://github.com/kubernetes/community/blob/master/contributors/devel/sig-architecture/api-conventions.md
20. Chen W et al (2020) Automated rolling updates in Kubernetes deployments. In: IEEE international conference on cloud engineering. Strategies for zero-downtime deployments. IEEE, pp 45–52
21. Kubernetes SIG-Network (2023) Kubernetes networking: services and beyond. Official documentation on Services and ClusterIP/NodePort. https://kubernetes.io/docs/concepts/services-networking/
22. CNCF Security TAG (2021) Managing configuration and secrets in Kubernetes. Tech. rep. Best practices for ConfigMaps and Secrets. https://github.com/cncf/tag-security/blob/main/guides/secret-management.md
23. Wells S (2021) Kubernetes YAML: from basics to advanced. Comprehensive guide to YAML structures for Kubernetes manifests. Manning, Shelter Island
24. Weaveworks (2022) GitOps for Kubernetes operators. Integrating operators with GitOps workflows. https://www.weave.works/blog/gitops-for-kubernetes-operators
25. Red Hat (2022) The operator framework: automating Kubernetes applications. Official framework for building Kubernetes Operators. https://operatorframework.io/
26. Łowicki M et al (2021) Design patterns for Kubernetes operators. In: ACM symposium on cloud computing. Case studies on PostgreSQL and Cassandra Operators. ACM, pp 312–327
27. Operator Framework (2023) Operator SDK: building Kubernetes operators. Toolkit for creating operators with Go/Ansible/Helm. https://sdk.operatorframework.io/
28. Aqua Security (2022) Troubleshooting Kubernetes: debugging pods and deployments. In: Practical debugging techniques for common issues. https://blog.aquasec.com/kubernetes-debugging-pods
29. Burns B, Beda J, Hightower K (2023) Kubernetes: Up and Running - Dive into the Future of Infrastructure, 3rd edn. Comprehensive guide to Kubernetes architecture and components. O'Reilly, Santa Rosa
30. Kubernetes Documentation (2023) Kubernetes components. Official documentation of Kubernetes architecture. https://kubernetes.io/docs/concepts/overview/components/
31. Huang W et al (2022) The Kubernetes scheduler: advanced techniques. In: KubeCon NA. Scheduling algorithms and custom schedulers. CNCF
32. Li X et al (2021) etcd: The definitive guide. Practical guide to etcd operations in Kubernetes. Packt
33. Chen D et al (2021) Kubelet: the Kubernetes node agent. In: USENIX symposium on cloud computing. Deep dive into Kubelet architecture. USENIX, pp 45–58
34. Xing Y et al (2021) Scaling Kubernetes to 5,000 nodes. In: ACM symposium on cloud computing. Empirical study on Kubernetes scalability limits. ACM, pp 407–419
35. AWS Containers Team (2021) Optimizing etcd for large-scale Kubernetes clusters. In: Performance tuning for etcd. https://aws.amazon.com/blogs/containers/optimizing-etcd-for-large-kubernetes-clusters/
36. VMware Tanzu Team (2021) Kubernetes controller design patterns. Tech. rep. How controllers maintain desired state. https://tanzu.vmware.com/content/blog/kubernetes-controller-design-patterns

37. Ongaro D, Ousterhout J (2014) The Raft consensus algorithm in etcd. In: USENIX ATC . Foundational paper on Raft consensus used in etcd. https://raft.github.io/raft.pdf
38. Red Hat OpenShift Team (2022) High availability Kubernetes clusters. In: Best practices for HA control planes. https://cloud.redhat.com/blog/designing-high-availability-kubernetes-clusters
39. Hockin T et al (2022) Kubernetes networking under the hood. In: CNCF Networking WG. How kube-proxy and CNI plugins work. https://github.com/kubernetes/community/tree/master/sig-network
40. Linux Foundation (2023) Kubernetes cluster operations. Operational guide for cluster administrators. https://training.linuxfoundation.org/kubernetes-cluster-ops/
41. Hausenblas M, Rice L (2021) Kubernetes vs. Docker Swarm: a comparative analysis. Comprehensive comparison of orchestration tools. O'Reilly, Santa Rosa
42. Docker Inc. (2022) Docker Swarm Mode: architecture and performance. Official documentation on Swarm's design. https://docs.docker.com/engine/swarm/how-swarm-mode-works/
43. Palma J et al (2020) Docker swarm for rapid deployment. In: IEEE international conference on cloud engineering. Case studies on Swarm's ease of use. IEEE, pp 112–125
44. Kubernetes SIG-Cluster-Lifecycle (2023) Simplifying Kubernetes installation with kubeadm. Cluster setup best practices. https://kubernetes.io/docs/setup/production-environment/tools/kubeadm/
45. CNCF Security TAG (2021) Kubernetes dashboard: security and RBAC. Tech. rep. Official dashboard documentation. https://github.com/kubernetes/dashboard
46. Tune E et al (2021) RBAC in Kubernetes: design and best practices. In: USENIX security symposium. RBAC implementation and security analysis. USENIX, pp 1023–1039
47. Kubernetes SIG-Auth (2023) Kubernetes RBAC documentation. Official RBAC guide for cluster access control. https://kubernetes.io/docs/reference/access-authn-authz/rbac/
48. Rice L, Hausenblas M (2022) Kubernetes security: operating RBAC and network policies. Practical RBAC and security policies. O'Reilly, Santa Rosa
49. Cresswell N (2022) Portainer for Docker swarm management. GUI-based Swarm administration. Leanpub
50. Wielgus M et al (2020) Autoscaling Kubernetes pods with horizontal pod autoscaler. In: IEEE international conference on cloud engineering. Algorithmic details of HPA's reactive scaling. IEEE, pp 201–215
51. Kubernetes SIG-Autoscaling (2023) Kubernetes HPA: official documentation. Configuration guide for HPA with CPU/custom metrics. https://kubernetes.io/docs/tasks/run-application/horizontal-pod-autoscale/
52. Google Cloud Team (2022) Dynamic node provisioning in Kubernetes. Google Cloud Architecture Center. Cloud provider integration with Cluster Autoscaler. https://cloud.google.com/kubernetes-engine/docs/concepts/cluster-autoscaler
53. Kubernetes SIG-Network (2023) Kubernetes load balancing: services and ingress. Official guide on traffic distribution. https://kubernetes.io/docs/concepts/services-networking/
54. Mirantis (2020) Docker Swarm Routing Mesh: design and implementation. Tech. rep. Deep dive into Swarm's built-in load balancing. https://www.mirantis.com/blog/docker-swarm-routing-mesh/
55. Ali S, Au M (2022) Kubernetes storage: persistent volumes in practice. Stateful workload management in Kubernetes. Packt
56. Portainer Team (2021) Docker swarm volume management. In: Best practices for storage in Swarm. https://www.portainer.io/blog/docker-swarm-volumes
57. Elastic (2023) Monitoring Docker swarm with cAdvisor and ELK. Third-party monitoring solutions for Swarm. https://www.elastic.co/blog/monitoring-docker-swarm
58. Bannister T, Liggitt J (2022) Understanding Kubernetes versioning: release cycles and compatibility. Kubernetes documentation. https://kubernetes.io/docs/setup/release/version-skew-policy/
59. Beda J, Burns B (2020) The evolution of Kubernetes: from v1.0 to v1.20. Cloud Native Computing Foundation (CNCF). https://www.cncf.io/blog/2020/12/07/kubernetes-1-20-evolution/

60. Kubernetes Storage SIG (2021) Persistent volume evolution in Kubernetes. Kubernetes storage documentation. https://kubernetes.io/docs/concepts/storage/persistent-volumes/
61. Kubernetes networking SIG (2022) The evolution of Kubernetes networking API: CNI and beyond. Kubernetes Blog. https://kubernetes.io/blog/2022/05/18/kubernetes-networking-evolution/
62. Brown M, Estes P (2021) Deprecation of Docker runtime and the transition to containerd. Kubernetes Blog. https://kubernetes.io/blog/2021/12/10/dockershim-faq/
63. Kubernetes SIG Node (2023) Dynamic resource allocation for GPUs and FPGAs in Kubernetes. Kubernetes enhancements. https://kubernetes.io/docs/concepts/scheduling-eviction/device-plugin/
64. Hockin T, Oppenheimer D (2023) Stabilization of sidecar containers in Kubernetes workloads. In: Kubernetes enhancement proposals (KEP). https://github.com/kubernetes/enhancements/tree/master/keps/sig-node/sidecar-containers
65. Kubernetes SIG Network (2023) Gateway API vs ingress: the future of Kubernetes traffic routing. Kubernetes Blog. https://kubernetes.io/blog/2023/06/15/gateway-api-progress/
66. Rice L, Coldwater I (2023) Kubernetes security evolution: trends and best practices. Cloud native security Blog. https://sysdig.com/blog/kubernetes-security-best-practices/
67. Google Cloud Kubernetes Team (2022) Performance benchmarking Kubernetes: scaling limits and metrics. Google Cloud Blog. https://cloud.google.com/blog/topics/containers-kubernetes/kubernetes-performance-benchmarking
68. Kubernetes Release Team (2023) Kubernetes upgrade path: best practices for smooth transitions. Kubernetes documentation. https://kubernetes.io/releases/version-skew-policy/
69. Cloud Native Computing Foundation (CNCF) (2023) CSI, CNI, and CRI interface version dependencies in Kubernetes. CNCF Technical Reports. https://github.com/container-storage-interface/spec
70. Helm and Kubernetes Operator SIG (2022) Helm and Kubernetes operator SDK compatibility. Helm Blog. https://helm.sh/blog/operator-sdk-compatibility/
71. Kubernetes Cloud Provider SIG (2023) Cloud provider support timelines for Kubernetes versions. Kubernetes cloud provider documentation. https://kubernetes.io/docs/concepts/architecture/cloud-controller/
72. Kubernetes SIG Architecture (2024) The Kubernetes future roadmap: features planned for v1.31+. Kubernetes enhancements. https://github.com/kubernetes/enhancements/tree/master/keps/sig-architecture
73. Liu W et al (2021) Scaling stateful applications in Kubernetes. IEEE Trans Parallel Distrib Syst
74. Xu R et al (2019) Stateful container orchestration in Kubernetes. IEEE Trans Cloud Comput
75. Bernstein D (2014) Containers and cloud: from LXC to docker to Kubernetes. IEEE Cloud Comput 1(3):81–84
76. Morabito R (2017) Performance evaluation of container-based virtualization for high performance computing environments. In: IEEE EUC
77. Merkel D (2014) Docker: lightweight Linux containers for consistent development and deployment. Linux J 2014(239):2
78. Sharma P et al (2016) Evaluation of docker containers for scientific workloads in the cloud. IEEE Trans Cloud Comput
79. Open Container Initiative (2015) Open container initiative runtime specification. Tech. rep. Linux Foundation
80. Burns B, Beda J, Hightower K (2019) Kubernetes: up and running. O'Reilly Media, Sebastopol
81. Luszczak M (2020) Kubernetes for developers. Packt Publishing, Birmingham
82. Hightower K, Burns B, Beda J (2017) Kubernetes in action. Manning Publications, Shelter Island
83. Kubernetes Authors (2023) Kubernetes Documentation: Persistent Volumes. Tech. rep. Cloud Native Computing Foundation
84. Abadi D et al (2020) Persistent data management in Kubernetes. In: ACM SIGMOD

85. Patel R et al (2023) Kubernetes operators for stateful services. In: USENIX OPED
86. Butler B (2022) Kubernetes security and observability. O'Reilly Media, Sebastopol
87. Kim S et al (2023) CSI drivers for Kubernetes: a survey. ACM Comput Surv
88. Garcia M et al (2023) kubectl best practices for cluster management. KubeCon EU
89. Kubernetes SIG-Apps (2023) Managing configuration data in Kubernetes. https://kubernetes.io/docs/concepts/configuration/
90. Rodriguez C (2023) Multi-cloud stateful Kubernetes deployments. J Distrib Syst
91. Thomas S (2023) Security implications of persistent storage in Kubernetes. IEEE Secur Priv
92. Google Cloud (2023) GKE persistent disk optimization. Tech. rep. Google
93. Zhang M et al (2022) StatefulSets vs. deployments: a comparative study. J Cloud Comput
94. White R (2023) Kubernetes patterns for stateful applications. Apress, New York
95. Hightower K, Burns B, Beda J (2021) Kubernetes up running: dive into the future of infrastructure, 3rd edn. O'Reilly Media, Sebastopol
96. Verma A et al (2020) Large-scale cluster management at Google with Borg. ACM Trans Comput Syst 34(4):1–47. https://doi.org/10.1145/2893170
97. Morabito R et al (2018) Evaluating performance of containerized applications in fog computing scenarios. IEEE Access 6:67245–67257. https://doi.org/10.1109/ACCESS.2018.2877659
98. Rizvi S, Rahman M, Mahfuz S (2021) Network performance evaluation of Kubernetes container orchestration platform. IEEE Trans Netw Serv Manag 18(2):1555–1571. https://doi.org/10.1109/TNSM.2021.3063378
99. Kumar B et al (2023) Optimal cloudlet selection in edge computing for resource allocation. SN Comput Sci 4(6):745
100. Verma VR et al (2025) An extensive investigation on Lyapunov optimization-based task offloading techniques in multi-access edge computing. SN Comput Sci 6(6):603
101. Liu F, Gupta I, Arpaci-Dusseau AC (2022) Optimizing disk I/O in containerized cloud environments. J Cloud Comput 11(1):1–22. https://doi.org/10.1186/s13677-022-00291-5
102. Kumar B et al (2025) Optimizing resource allocation in cloud-native applications through proactive autoscaling with the InformerAutoScale model. J Supercomput 81(9):1077
103. Hwang J, Kim J, Kim TH (2019) Performance monitoring and analysis of cloud-native applications. IEEE Cloud Comput 6(3):60–69. https://doi.org/10.1109/MCC.2019.2916448
104. Mao M, Humphrey M (2020) Auto-scaling to minimize cost and meet application deadlines in cloud workflows. Future Gener Comput Syst 72:205–218. https://doi.org/10.1016/j.future.2016.06.015
105. Pahl C et al (2023) Proactive auto-scaling and predictive resource allocation in Kubernetes environments. ACM Comput Surv 55(1):1–32. https://doi.org/10.1145/3506548
106. Abdullah M et al (2022) Predictive autoscaling for microservices using reinforcement learning. IEEE Trans Cloud Comput
107. Chen W, et al (2020) Proactive autoscaling in Kubernetes using LSTM networks. J Cloud Comput 9(1):1–18
108. Wang L, et al (2022) Hybrid autoscaling in Kubernetes: combining reactive and proactive approaches. Future Gener Comput Syst 128:1–15
109. Li Y, et al (2021) Transformer-based workload forecasting for cloud autoscaling. In: IEEE international conference on cloud engineering (IC2E)
110. Sharma P, et al (2021) InformerAutoScale: a scalable transformer for cloud workload forecasting. In: ACM symposium on cloud computing (SoCC)
111. Kumar B, Verma A, Verma P (2024) Optimizing resource allocation using proactive scaling with predictive models and custom resources. Comput Electr Eng 118:109419
112. Tan W, et al (2023) Comparative analysis of reactive vs. proactive autoscaling in cloud. IEEE Cloud Comput
113. Fernandez C, et al (2023) Serverless autoscaling with proactive reinforcement learning. ACM Trans Auton Adapt Syst
114. Kumar R, et al (2021) Reinforcement learning for dynamic resource allocation in Kubernetes. IEEE Trans Parallel Distrib Syst

115. Nguyen T, et al (2022) Deep reinforcement learning for Kubernetes cluster autoscaling. In: IEEE International conference on cluster computing (CLUSTER)
116. Liu X, et al (2021) Bi-LSTM for predictive autoscaling in cloud-native systems. IEEE Access 9:123456–123467
117. Zhang H, et al (2020) ARIMA-based predictive autoscaling for cloud applications. In: IEEE International conference on autonomic computing (ICAC)
118. Li Y, et al (2023) Proactive autoscaling in Kubernetes using LSTM and MAPEK. In: IEEE International conference on autonomic computing (ICAC), pp 45–54
119. Yang L, et al (2021) XGBoost for predictive autoscaling in cloud environments. In: IEEE International conference on big data (big data)
120. Singh A, et al (2023) MAPE-K framework for intelligent autoscaling in Kubernetes. In: IEEE International conference on cloud computing (CLOUD)
121. Kumar B, Verma A, Verma P (2025) A multivariate transformer-based monitor-analyze-plan-execute (MAPE) autoscaling framework for dynamic resource allocation in cloud environment. Computing 107(3):69
122. Chen W, et al (2022) MAPE-K for autonomous cloud resource management: a survey. IEEE Trans Cloud Comput 10(2):987–1002
123. Nguyen T, et al (2024) MV-transformer: long-term workload forecasting for autoscaling. IEEE Trans Parallel Distrib Syst
124. Kubernetes Authors (2023) Kubernetes official documentation: installation. https://kubernetes.io/docs/setup/
125. Patel R, et al (2022) Cost-efficient autoscaling using deep Q-learning in Kubernetes. J Netw Syst Manag
126. Prometheus Authors (2023) Prometheus monitoring system. https://prometheus.io/docs/introduction/overview/
127. IBM Research (2023) IBM edge application manager: MAPE-K for edge autoscaling. IBM J Res Dev 67(3):5:1–5:15
128. Docker Inc. (2023) Docker installation guide. https://docs.docker.com/engine/install/
129. Gupta A, et al (2022) MAPE-K for IoT workloads: edge-cloud autoscaling. IEEE Internet Things J 9(18):17562–17576
130. Google Cloud AI Team (2022) AI-driven autoscaling in Google cloud using MAPE-K. In: IEEE International conference on cloud engineering (IC2E), pp 33–44
131. Nokia Bell Labs (2024) Nokia AVA: MAPE-K for 5G network autoscaling. IEEE Trans Netw Serv Manag
132. Singh A, et al (2023) IBM Watson AIOps: self-healing via MAPE-K. In: IEEE/I-FIP network operations and management symposium (NOMS), pp 1–9
133. Zhang H, et al (2023) Reinforcement learning for MAPE-K-based autoscaling. In: ACM/IEEE symposium on edge computing (SEC), pp 78–92
134. Fernandez C, et al (2023) Meta-learning for adaptive MAPE-K autoscaling. In: IEEE International conference on cloud computing (CLOUD), pp 311–325
135. Tan W, et al (2024) Energy-efficient autoscaling using MAPE-K and DRL. Sustain Comput 41:100943
136. Kumar R, et al (2023) Digital twins for MAPE-K-based autoscaling. In: IEEE International conference on autonomic computing, pp 201–210
137. Burns B, et al (2016) Borg, omega, and Kubernetes: lessons learned from three container-management systems over a decade. Commun ACM 59(5):50–57
138. Lorido-Botran T, Miguel-Alonso J, Lozano JA (2014) A review of auto-scaling techniques for elastic applications in cloud environments. J Grid Comput 12(4):559–592
139. Guleria K, Jain A, Shukla S (2021) Autoscaling in cloud computing: a survey. J Netw Comput Appl 185:103080
140. Khazaei H, Xiao W, Lin M (2020) A survey of autoscaling in cloud computing. IEEE Trans Serv Comput 14(1):88–103

141. Yao H, et al (2022) Informer: beyond efficient transformer for long sequence time-series forecasting. In: Proceedings of the AAAI conference on artificial intelligence, vol 35(12), pp 11106–11115
142. Cheng Y, Li K, Wang X (2019) Autoscaling web applications in cloud using machine learning: a systematic review. IEEE Access 7:112292–112307
143. Chen X, Zhang H, Wang J (2020) Autoscaler: a proactive scaling approach for microservice applications in cloud environments. IEEE Trans Cloud Comput 8(1):223–236
144. Zhou H, et al (2021) Informer: beyond efficient transformer for long sequence time-series forecasting. In: Proceedings of the AAAI conference on artificial intelligence, vol 35(12), pp 11106–11115
145. Lammel R (2004) NASA HTTP web server logs. https://ita.ee.lbl.gov/html/contrib/NASA-HTTP.html. Accessed from NASA Kennedy Space Center Web Server Log dataset
146. Zhang J, Zheng Z, Zhang Y (2016) Deep learning for predictive analytics: a survey. IEEE Access 4:1374–1389
147. Wang H, Liu Y, Liu Y (2020) Deep learning for load forecasting with smart meter data: a review. Energy Rep 6:1256–1274
148. Xu X, Bao Y, Liu Y (2021) A survey of machine learning techniques for workload prediction in cloud environments. ACM Comput Surv 54(7):1–36
149. Bergmeir C, Benitez JM (2018) A note on the validity of cross-validation for evaluating time series prediction. Expert Syst Appl 45:208–210
150. Kingma DP, Ba J (2014) Adam: a method for stochastic optimization. arXiv preprint arXiv:1412.6980
151. Halkidi M, Mitkas PA (2021) Predictive autoscaling for containerized applications using LSTM neural networks. Future Gener Comput Syst 117:86–97
152. Mao H, et al (2016) Resource management with deep reinforcement learning. In: Proceedings of the 15th ACM workshop on hot topics in networks, pp 50–56
153. Morabito R (2018) Evaluating performance and energy efficiency of containerized applications in Kubernetes clusters. Cluster Comput 21(1):947–962
154. Kim Y, Lee J, Park S (2021) A comparison of deep learning models for forecasting network traffic. In: 2021 IEEE International conference on big data and smart computing (BigComp). IEEE, pp 347–350
155. Tang H, Zhang J, Liu Y (2023) An intelligent container autoscaler using prediction and anomaly detection for cloud applications. Future Gener Comput Syst 138:142–157
156. Li Q, et al (2022) Autoscaling cloud-native applications with service-level objective guarantees. IEEE Trans Parallel Distrib Syst 33(10):2325–2340
157. Zhu Y, Qian C, Guo M (2020) Learning-based dynamic resource allocation for containerized cloud applications. IEEE Trans Serv Comput 13(3):543–556
158. Gupta A, Jayachandran P, Rao P (2020) A survey of reinforcement learning techniques for auto-scaling in cloud computing. J Syst Softw 168:110670
159. Kubernetes Community (2023) Minikube getting started. https://minikube.sigs.k8s.io/docs/start/
160. CIS Benchmarks (2023) Kubernetes security and hardening guide. Tech. rep. Security best practices for Kubernetes components. https://www.cisecurity.org/benchmark/kubernetes/
161. Red Hat (2023) Installing Kubernetes on CentOS. https://access.redhat.com/documentation/en-us/openshift_container_platform
162. Canonical (2023) Kubernetes on Ubuntu: official guide. https://ubuntu.com/kubernetes/docs
163. Tigera Inc. (2023) Calico networking for Kubernetes: performance and security. Calico Documentation. https://docs.projectcalico.org/
164. Isovalent (2023) Cilium for Kubernetes networking. https://docs.cilium.io/en/stable/gettingstarted/k8s-install-default/
165. Kubernetes SIG-UI (2023) Kubernetes dashboard setup. https://github.com/kubernetes/dashboard
166. Prometheus Authors (2023) Monitoring Kubernetes with prometheus. https://prometheus.io/docs/guides/k8s/

167. HashiCorp (2023) Automating Kubernetes with Terraform. https://developer.hashicorp.com/terraform/tutorials/kubernetes
168. Patel R, et al (2021) Infrastructure as code for Kubernetes: Terraform vs. Ansible. In: ACM symposium on cloud computing (SoCC), pp 311–325
169. NSA/CISA (2021) Role-based access control in Kubernetes: security best practices. In: Kubernetes hardening guide. https://media.defense.go/2021/Aug/03/2002820425/-1/-1/1/CTR_KUBERNETES%20HARDENING%20GUIDANCE.PDF
170. Burns B, et al (2020) Troubleshooting Kubernetes: a systematic approach. In: USENIX symposium on cloud computing
171. Kubernetes Authors (2023) Kubernetes documentation: deployments. https://kubernetes.io/docs/concepts/workloads/controllers/deployment/
172. Google Cloud Team (2022) Declarative vs. imperative Kubernetes management. In: KubeCon NA
173. Microsoft Azure Team (2023) Kubernetes deployment anti-patterns. Azure Architecture Center. https://learn.microsoft.com/en-us/azure/architecture/guide/kubernetes/anti-patterns
174. Kubernetes SIG-CLI (2023) kubectl command reference. https://kubernetes.io/docs/reference/generated/kubectl/kubectl-commands
175. Burns B, et al (2021) Kubernetes YAML patterns: structure and reusability. ACM Queue 19(3)
176. Wang L, et al (2022) Autoscaling in Kubernetes: a comparative study. In: IEEE International conference on cloud engineering (IC2E), pp 45–58
177. Kubernetes SIG-Autoscaling (2023) Horizontal pod autoscaler walkthrough. https://kubernetes.io/docs/tasks/run-application/horizontal-pod-autoscale/
178. Chen W, et al (2023) Zero-downtime deployments in Kubernetes: strategies and trade-offs. IEEE Trans Cloud Comput 11(2)
179. NGINX Inc. (2023) Deploying NGINX in Kubernetes. https://docs.nginx.com/nginx-ingress-controller/installation/installation-with-manifests/
180. Sharma P, et al (2021) Efficient resource allocation for containerized workloads. In: ACM symposium on cloud computing (SoCC), pp 201–215
181. Kubernetes SIG-UI (2023) Kubernetes dashboard user guide. https://github.com/kubernetes/dashboard
182. Kubernetes Authors (2023) Kubernetes storage concepts. https://kubernetes.io/docs/concepts/storage/
183. Kubernetes SIG-Storage (2023) Container storage interface (CSI) specification. https://kubernetes-csi.github.io/docs/
184. Cloud Native Computing Foundation (2023) Persistent volumes in practice. https://github.com/kubernetes/examples/tree/master/staging/volumes
185. Li Y, et al (2022) Comparative analysis of Kubernetes storage backends. IEEE Trans Cloud Comput 10(3):1450–1464
186. Burns B, et al ((2022)) Patterns for stateful applications in Kubernetes. ACM Queue 20(2)
187. F5 Networks (2023) NGINX ingress controller documentation. https://docs.nginx.com/nginx-ingress-controller/
188. NSA/CISA (2022) Zero trust networking in Kubernetes. In: Kubernetes hardening guide. https://media.defense.gov/2022/Aug/29/2003066362/-1/-1/1/CTR_KUBERNETES_HARDENING_GUIDE.PDF.
189. IBM Research (2023) Scalability limits of Kubernetes storage. J Cloud Comput 12(1)
190. Istio Authors (2022) Istio security for Kubernetes microservices. IEEE Cloud Comput 8(3)
191. Let's Encrypt (2023) Securing Kubernetes ingress with TLS. https://letsencrypt.org/docs/certificates-for-kubernetes/
192. Google Cloud Team (2020) RBAC for Kubernetes: design and implementation. In: USENIX security symposium
193. Kubernetes Authors (2023) Kubernetes security overview. https://kubernetes.io/docs/concepts/security/
194. Sysdig (2023) Falco for Kubernetes runtime security. J Cloud Secur 5(1)

195. NSA/CISA (2022) Implementing zero trust in Kubernetes clusters. Cybersecurity Advisory https://media.defense.gov/2022/Aug/29/2003066362/-1/-1/1/CTR_KUBERNETES_HARDENING_GUIDE.PDF
196. Aqua Security (2023) Trivy vulnerability scanner. https://aquasecurity.github.io/trivy/
197. Prometheus Authors (2023) Monitoring Kubernetes with prometheus. https://prometheus.io/docs/guides/k8s/
198. Kubernetes SIG-UI (2023) Securing Kubernetes dashboard. https://github.com/kubernetes/dashboard/wiki/Access-control
199. Cloud Native Computing Foundation (2022) OIDC authentication in Kubernetes. In: KubeCon EU
200. HashiCorp (2023) Kubernetes secrets best practices. https://developer.hashicorp.com/vault/docs/platform/k8s
201. Grafana Labs (2022) Grafana for Kubernetes: advanced monitoring techniques. In: KubeCon NA
202. CNCF (2023) EFK stack for Kubernetes logging. https://github.com/kubernetes-sigs/instrumentation-addons
203. RedLock Research Team (2019) Cryptojacking incident analysis: Tesla Kubernetes breach. In: IEEE symposium on security and privacy, pp 112–125
204. Red Hat (2023) Keycloak operator for Kubernetes. https://www.keycloak.org/getting-started/getting-started-kubernetes
205. Microsoft Azure Team (2023) Secure CI/CD pipelines for Kubernetes. Azure Architecture Center. https://learn.microsoft.com/en-us/azure/architecture/guide/kubernetes/secure-ci-cd
206. Google SRE Team (2021) Kubernetes admission controllers: security best practices. In: USENIX security symposium
207. Kubernetes SIG-Security (2023) Pod security standards. https://kubernetes.io/docs/concepts/security/pod-security-standards/
208. Google Cloud (2023) Serverless workloads with Knative. CloudNativeCon
209. Improbable (2022) Cross-cluster monitoring with Thanos. In: PromCon
210. Kubernetes SIG-Multicluster 2023 Kubernetes Federation v2 Documentation. https://github.com/kubernetes-sigs/kubefed
211. Google Cloud Team (2022) Design patterns for multi-cluster kubernetes. In: KubeCon NA
212. Suen J, et al (2022) ArgoCD: declarative GitOps for Kubernetes. CNCF Project. Tool for GitOps-based Kubernetes deployments. https://argoproj.github.io/argo-cd/
213. Weaveworks (2022) GitOps: continuous deployment for Kubernetes. Commun ACM 65(6)
214. Red Hat (2021) Kubernetes operators: design and implementation. J Cloud Native 5(2)
215. Niantic Engineering (2017) Scaling Pokémon Go on Kubernetes. In: SRECon
216. Booking.com Tech Blog (2022) Booking.com's Kubernetes migration journey. In: ACM Queue 20(4)
217. AWS EKS Team (2021) High-availability Kubernetes control plane. In: USENIX ATC
218. CNCF (2023) Helm: the package manager for Kubernetes. https://helm.sh/docs/
219. Microsoft Research (2023) AI/ML orchestration in Kubernetes. IEEE Cloud Comput 9(3)
220. Kubeflow Authors (2023) Kubeflow: machine learning on Kubernetes. https://www.kubeflow.org/
221. Linux Foundation (2023) Cloud native career paths. In: Open source summit
222. Cloud Native Computing Foundation (2023) CKA exam curriculum. https://github.com/cncf/curriculum
223. CNCF (2023) Kubernetes community resources. https://kubernetes.io/community/
224. Linux Foundation (2023) CKAD certification study guide. https://training.linuxfoundation.org/certification/certified-kubernetes-application-developer-ckad/
225. Kubernetes.io (2023) Kubernetes interview questions. Official Documentation. https://kubernetes.io/docs/reference/glossary/

FSC
www.fsc.org
MIX
Papier aus verantwortungsvollen Quellen
Paper from responsible sources
FSC® C105338

If you have any concerns about our products,
you can contact us on
ProductSafety@springernature.com

In case Publisher is established outside the EU,
the EU authorized representative is:
Springer Nature Customer Service Center GmbH
Europaplatz 3, 69115 Heidelberg, Germany

Printed by Libri Plureos GmbH
in Hamburg, Germany